Women in the History of Philosophy and Sciences

Volume 23

As the historical records prove, women have long been creating original contributions to philosophy. We have valuable writings from female philosophers from Antiquity and the Middle Ages, and a continuous tradition from the Renaissance to today. The history of women philosophers thus stretches back as far as the history of philosophy itself. The presence as well as the absence of women philosophers throughout the course of history parallels the history of philosophy as a whole.

Edith Stein, Hannah Arendt and Simone de Beauvoir, the most famous representatives of this tradition in the twentieth century, did not appear from nowhere. They stand, so to speak, on the shoulders of the female titans who came before them.

The series Women Philosophers and Scientists published by Springer is of interest not only to the international philosophy community, but also for scholars in history of science and mathematics, the history of ideas, and in women's studies.

Clara Carus · Jeffrey K. McDonough
Editors

Émilie Du Châtelet in Relation to Leibniz and Wolff

Similarities and Differences

Springer

Editors
Clara Carus
Department of Philosophy
Harvard University
Cambridge, MA, USA

Jeffrey K. McDonough
Department of Philosophy
Harvard University
Cambridge, MA, USA

ISSN 2523-8760 ISSN 2523-8779 (electronic)
Women in the History of Philosophy and Sciences
ISBN 978-3-031-80543-1 ISBN 978-3-031-80541-7 (eBook)
https://doi.org/10.1007/978-3-031-80541-7

This Springer imprint is published by the registered company Springer Nature Switzerland AG
The registered company address is: Gewerbestrasse 11, 6330 Cham, Switzerland

Series Foreword

Women in the History of Philosophy and Sciences

The history of women's contributions to philosophy and the sciences dates back to the very beginnings of these disciplines. Theano, Hypatia, Du Châtelet, Agnesi, Germain, Lovelace, Stebbing, Curie, and Stein are only a small selection of prominent women philosophers and scientists throughout history.

The Springer Series *Women in the History of Philosophy and Sciences* provides a platform for publishing cutting-edge scholarship on women's contributions to the sciences, to philosophy, and to interdisciplinary academic areas. We therefore include in our scope women's contributions to biology, physics, chemistry, and related sciences. The series also encompasses the entire discipline of the history of philosophy since antiquity (including metaphysics, aesthetics, and philosophy of religion). We welcome also work about women's contributions to mathematics and to interdisciplinary areas such as philosophy of biology, philosophy of medicine, and sociology.

The research presented in this series serves to recover women's contributions and to revise our knowledge of the development of philosophical and scientific disciplines, so as to present the full scope of their theoretical and methodological traditions. Supported by an advisory board of internationally esteemed scholars, the volumes offer a comprehensive, up-to-date source of reference for this field of growing relevance. See the listing of planned volumes.

The Springer Series *Women in the History of Philosophy and Sciences* will publish monographs, handbooks, collections, anthologies, and dissertations.

Paderborn, Germany — Ruth Hagengruber
Cleveland, USA — Mary Ellen Waithe
Vercelli, Italy — Gianni Paganini

Series Foreword

Acknowledgements

The present volume is a result of the research project *Émilie Du Châtelet in Relation to Leibniz and Wolff—Similarities and Differences*, which was generously funded by the German Research Foundation (Reference Number: 471747296). As part of the project, Jeffrey K. McDonough was named Mercator Fellow and spent one year at Paderborn University, enabling close collaboration on the topic. Besides the German Research Foundation, the editors wish to express their gratitude to Ruth E. Hagengruber, director of the project, for her hands-on and dedicated support as well as to the team at the Center for the History of Women Philosophers, directed by Ruth E. Hagengruber, for creating a wonderful research environment. We also wish to thank Harvard University, which supported the Harvard History of Philosophy Workshops at Paderborn University during the duration of the project. We wish to thank all the speakers and participants of the conference *Émilie Du Châtelet in Relation to Leibniz and Wolff—Similarities and Differences* in September 2022 in Paderborn. Last but not least, we wish to thank the contributors of this volume for their scholarship, insights, and unwavering dedication.

Contents

Editors and Contributors

About the Editors

Clara Carus is Alexander-von-Humboldt Postdoctoral Fellow at Harvard University. Her research focus is on Early Modern Philosophy, especially the works of Du Châtelet and Leibniz.

Jeffrey K. McDonough is Professor of Philosophy at Harvard University. His primary areas of research lie in early modern philosophy, the philosophy of science, and especially their intersection.

Contributors

Fatema Amijee University of British Columbia, Vancouver, Canada

Julia Borcherding University of Cambridge, Cambridge, UK

Clara Carus Department of Philosophy, Harvard University, Cambridge, MA, USA

Katherine Dunlop Department of Philosophy, University of Texas at Austin, Austin, TX, USA

Jan J. Forsman University of Iowa, Iowa City, IA, USA

Jani I. Hakkarainen Tampere University, Tampere, Finland

Qiu Lin Simon Fraser University, Burnaby, BC, Canada

Jeffrey K. McDonough Department of Philosophy, Harvard University, Cambridge, MA, USA

Gabriel Müller University Library Basel, Basel, Switzerland

Anne-Lise Rey Université Paris Nanterre, IRePh, Paris, France

Chapter 1
Introduction to *Émilie Du Châtelet in Relation to Leibniz and Wolff: Similarities and Differences*

Clara Carus and Jeffrey K. McDonough

Émilie Du Châtelet was an extraordinary person.[1] Born in Paris in 1706 into an aristocratic family, she received an excellent, comprehensive education. She learned Greek, Latin, Italian, and German. She studied mathematics, science, and literature. At the age of 18, she married the Marquis Florent-Claude de Châtelet-Lomont. Together they would have three children, the second dying shortly after birth. She adored the opera and theater, enjoyed parties and playing cards, but she was determined to acquire fame and glory through her philosophical and scientific achievements. She cultivated a broad network of correspondents, including Francesco Algarotti, Laura Bassi, Alexis Claude Clairaut, Leonhard Euler, Françoise de Graffigny, Frederik the Great, Jean-Jacques d'Ortous de Mairan, Pierre Louis Moreau de Maupertuis, Julien Offray de la Mettrie, and several members of the Bernoulli family. In 1733, she became romantically and intellectually involved with Voltaire.[2] Fleeing Voltaire's enemies, they moved to Du Châtelet's husband's estate, Cirey, in the Champagne region of northeastern France. Together they transformed the estate into an intellectual oasis. They hosted distinguished scientists, intellectuals, and writers. They added a laboratory and expanded the library. They studied. They experimented. They put on plays. By the late 1740's, however, Du Châtelet and Voltaire had drifted apart romantically. In 1748, Du Chatelet began a new affaire, this time with the younger

[1] For biographical treatments of Du Châtelet's life, see Badinter and Muzerelle (2006), Ehrman (1986), Zinsser (2006), Zinsser (2007), Detlefsen and Janiak (2018), Carus (2024b), Hagengruber (N.d.). Project Vox Team (2019).

[2] For a detailed study of Du Châtelet's relationship with Voltaire, see Wade (1941).

C. Carus · J. K. McDonough (✉)
Department of Philosophy, Harvard University, Cambridge, MA, USA
e-mail: jkmcdon@fas.harvard.edu

C. Carus
e-mail: claracarus@fas.harvard.edu

C. Carus and J. K. McDonough (eds.), *Émilie Du Châtelet in Relation to Leibniz and Wolff*, Women in the History of Philosophy and Sciences 23,
https://doi.org/10.1007/978-3-031-80541-7_1

French poet, author, and soldier Jean-François de Saint-Lambert. She soon became pregnant. Now aged 42, Du Chatelet was well-aware of the risks to her health. Fearing the worst, she worked tirelessly to secure her children's future and complete her intellectual projects. In September 1749, she gave birth to a daughter, but died, tragically, from complications six days later. Hers was a short, but undeniably extraordinary life.

Du Châtelet was also an accomplished scholar.[3] After arriving at Cirey, she undertook a free translation of Mandeville's controversial social satire, *Fable of the Bees*, adding a substantial preface of her own. She lent her support to—perhaps even co-authored—Voltaire's *Elements of the Philosophy of Newton.* She began a work on optics and carried out experiments on the nature of fire, secretly submitting an essay on the latter topic to the Royal Academy of Science's prize competition for 1738. She planned a work on grammar and dedicated years to a withering critique of the Bible.[4] In 1740, she published the first edition of her most philosophically influential work, her *Institutions de Physique* anonymously. A revised and expanded edition appeared in 1742, this time with her name in the title. She wrote an elegant, if slightly sad, essay on happiness. She penned a treatise on freedom, long attributed to Voltaire, and carried out a major translation of, and commentary on, Isaac Newton's *Principia Mathematica.* Her translation would become the standard French language version of Newton's masterpiece, and her commentary (as well as some important passages from the *Institutions*) would be copied verbatim (but without acknowledgment) in the famed Enlightenment *Encyclopédie* of Denis Diderot and Jean-Baptiste le Rond d'Alembert. Du Châtelet's oeuvre is currently the focus of intense scholarly attention, and her work is finding its way back into today's classrooms (see, Hagengruber, 2024). By any measure, Du Châtelet was a remarkably accomplished scholar.

An extraordinary person and an accomplished scholar, was Du Châtelet also an original philosopher? Although scholarship on Du Châtelet is still gaining momentum, thus far there has been little consensus. Proponents of what we might call *derivative interpretations* suggest that the answer is "no." On this view, Du Châtelet primarily repackaged the philosophical views of others, and had no original philosophy of her own. Such interpretations go back to Du Châtelet's own day. In 1739, Du Châtelet hired Samuel König, a former student of Christian Wolff, to tutor her in geometry. König had not been Du Châtelet's first choice, and their relationship quickly soured. Within the year, he was accusing her of plagiarizing his notes in her impending *Institutions de Physique* (Droysen, 1909, 232; Janik, 1982, 97; Reichenberger, 2016, 53ff). Although Du Châtelet immediately sought to defend herself (see, Kölving and Brown, 2018, [Letter 280] 1:568–9), many were willing to take König at his word, especially against a woman. Frederick II of Prussia—"Frederick the Great"—for example, initially found König's charge plausible. Writing to his advisor Charles-Étienne Jordan that "La Minerve [i.e., Du Châtelet] has just done

[3] For discussion of Du Châtelet's scholarly achievements, see Zinsser (2007), Hutton (2004), Reichenberger (2016).

[4] For Du Châtelet's *Examens de la Bible* see, Hagengruber (2020), Seguin (2021), Seguin (2022a, 2022b).

her physics," he remarks that "There is some good in it; it was Koenik who dictated her theme." (Preuss, 1851, [Vol. 17] 71) Wolff similarly saw his own thoughts in Du Châtelet's writings. Having read the first edition of Du Châtelet's *Institutions*, Wolff wrote to Count Ernst Christof von Manteuffel in May of 1641, to declare, "I am amazed that this lady can present things with such great clarity, and if she were to summarize my entire philosophy in the same manner as she promised, I do not doubt that I would have to recognize her as my apostle in France" (Middell & Neumann, 2013, Letter 101 [lines 13–19]). Later he would add, in what must have seemed to him to be high praise, "When she speaks about what I have presented in my metaphysics, it is as if I were hearing myself lecture" (Middell & Neumann, 2013, Letter 102 [14 June 1641, to Manteuffel]). Derivative interpretations, initiated in Du Châtelet's day, have persisted to our own. The twentieth century scholar William Barber, for example, has proposed that "in spite of her in other respects somewhat masterful personality," "Mme du Châtelet's scientific and philosophical ideas appear, on re-examination, to be essentially derivative [of Leibniz]. Her happiest role here is perhaps that of a disciple—of Voltaire, of Maupertuis, of König" (Barber, 1967, 221).

It seems not implausible that derivative interpretations have been encouraged, at least in some instances, by ego, jealousy, and sexist assumptions. Nonetheless, two considerations might be seen as lending prima facie support to the idea that Du Châtelet's philosophy is essentially derived from the philosophy of others. The first such consideration is that several of Du Châtelet's own statements straightforwardly imply that she saw herself as a disciple of the philosophy of Wolff and Leibniz. The most striking passage in this respect is Du Châtelet's declaration in the concluding paragraph of the Preface to the *Institutions*. There, she writes:

> Thus, in the first few chapters, I'll outline the principal opinions of G.W. Leibniz on Metaphysics, which I have taken from the works of the famous Christian Wolff, whom you have heard me talk so much about with one of his Disciples who stayed with me for some time and who sometimes gave me extracts from his work. (Inst1742, XII)

In this passage, Du Châtelet seems to tell us flat-out that the most philosophical chapters of the *Institutions* outline what she takes to be the principal opinions of Leibniz's metaphysics. She seems to indicate explicitly that her understanding of those principal opinions is derived from the works of Wolff and discussions with "one of his Disciples," the disciple in question being, of course, König himself. Scattered lines from Du Châtelet's correspondence may appear to confirm the impression that the quoted passage is prone to give. Thus, for example, in a letter dated April 1740, Du Châtelet tells Frederick that she felt obliged to write her *Institutions* for the benefit of her son because she became "convinced that physics cannot do without metaphysics," and thus "wanted to give him an idea of the metaphysics of Mr. Leibniz, which I admit is the only one that has satisfied me" (Kölving and Brown, 2018, [Letter 286] 1:576). Similarly, in another letter to Frederick, this one dated 11 August 1740, Du Châtelet writes, "I have the intention of presenting an entire philosophy in French, in the style of Wolff but with a French flavor," noting further that "Wolff's works would repel the French lightness by their form alone, but I am convinced that my compatriots will appreciate this precise and rigorous way of reasoning" (Kölving & Brown, 2018, [Letter 297] 1:592).

A second consideration lending prima facie support to derivative interpretations can be found in the historical context surrounding the publication of the *Institutions*.[5] Du Châtelet appears to have drafted the *Institutions* over the course of a year, completing a manuscript by the end of the summer of 1738. Seeking required approval from the royal government, she submitted a manuscript to Henri Pitot (Zinsser, 2006, 167). Pitot endorsed the work, notably describing it as having "presented the principles of philosophy of Mr. Leibnits [sic] & that of Mr. Newton" (Zinsser, 2006, 167). With Pitot's approval in hand, Du Châtelet promptly sent the manuscript to the French printer Pierre Prault, whose motto, fittingly enough, was *serere ne dubites*—"do not hesitate to sow." But Du Châtelet did hesitate. By the onset of winter, her dispute with Jean-Jacques d'Ortous de Mairan over the conservation of vis viva seems to have prompted her to engage more deeply with the views of Leibniz and Wolff. She appears to have carefully read, perhaps for the first time, the French translations of Wolff's writings that Frederick had previously sent to Voltaire,[6] and reportedly wrote to Wolff that "she wants to learn German in order to read [his] German writings."[7] Her earnest engagement with Wolff's writings, in turn, appears to have served as the impetus behind her decision to halt the already half-completed printing of the *Institutions* in order to develop its most philosophical themes, including its treatment of space, time, God, and the principles of knowledge (Zinsser, 2006, 170; see also Janik, 1982, 94). The historical context of the publication of the *Institutions* thus may appear to lend some credibility to derivative interpretations by suggesting that Du Châtelet sought to fill what she perceived as a lacuna in her 1738 manuscript with quickly-made summaries of philosophical views she gleaned from Leibniz, Wolff, and König.

The evidence in support of derivative interpretations is intriguing enough that it deserves to be taken seriously. But it is hardly conclusive, and the question of originality itself is more complex than derivative interpretations imply. What we might call *synthetic interpretations* propose that derivative interpretations overlook the true source of Du Châtelet's originality. They suggest that the innovation of Du Châtelet's philosophy is to be found precisely in the way in which it synthesizes the views of others to create something new and original. In developing this line of thought, it may be helpful to distinguish between what we might call philosophically *modest* and philosophically *ambitious* synthetic interpretations.

Philosophically modest synthetic interpretations see novelty in Du Châtelet's efforts to reconcile specifically Newtonian physics and Leibnizian metaphysics. They

[5] For discussion of the development of the *Institutions de Physique* see Zinsser 166–167, 169–177; see also Janik (1982) and Neumann et al. (N.d.).

[6] See the postscript of Wolff's letter to Manteuffel on 14 June 1741 (Middell and Neumann, 2013, Letter 102 [lines 64 to 65]). Du Châtelet herself refers to Wolff's Latin *Ontologia* in the preface of the *Institutions de Physiques* (Inst1742, XIIfn.).

[7] Wolff citing Du Châtelet in his letter to Manteuffel, dated 3 April 1740 (Middell and Neumann, 2013, Letter 60).

urge us to appreciate the creativity involved in bringing together—to put it roughly—British science and Continental philosophy. Given the nationalist divides that had separated science on the island and philosophy on the continent—divides that Du Châtelet explicitly and rightly bemoans (Inst1742, VII)—it should be conceded that her attempt to do justice to both was no mean task, and the extent to which she managed the feat, no small accomplishment. It might nonetheless be counted as philosophically modest for our purposes insofar as it is consistent with Du Châtelet's having no original philosophical views of her own. Linda Gardiner Janik's influential work might suggest such an interpretation. For Janik explicitly affirms the originality of Du Châtelet's thought, telling us that "the version of Newtonian physics given" by Du Châtelet constitutes a "radically new position," precisely because of the ways in which Du Châtelet reinterprets Newton's physics through the lens of Wolff's ontology. Furthermore, she unambiguously affirms that the "originality" of her "synthesizing project … can not be doubted" (see Janik, 1982, 98, 110). Nonetheless, Janik seems to suggest that Du Châtelet's philosophy per se is not original, declaring, for example, that "we cannot attribute any original metaphysical doctrines to mme Du Châtelet" (Janik, 1982, 103). At the risk of oversimplifying her rich view, Janik's reading thus suggests one way of seeing Du Chatelet as an original synthetic thinker without seeing her as developing any original philosophical doctrines per se.

Philosophically ambitious synthetic interpretations, in contrast, insist that Du Châtelet did not merely bring together science and philosophy but also wove together the philosophical views of others to form a novel philosophical outlook of her own. Here we can further distinguish two distinct, but compatible, philosophically ambitious synthetic projects.

The first philosophically ambitious synthetic project would involve Du Châtelet's drawing together threads from the Leibnizian resources available to her. Those resources include Leibniz's *Theodicy* and the Leibniz-Clarke correspondence in Pierre Desmaizeux's French translation (Bour and Zinsser, 2009, 156, fn 77). Du Châtelet also had access to many of Leibniz's published papers as well as replies to those papers, (see Besterman, 1958 1[D1636, 24 October 1738]; Janik, 1982, 89). On this point it is worth emphasizing what a limited view of Leibniz's works these resources would have given Du Châtelet. For Leibniz never published the vast majority of his writings. Indeed, he once confided to a correspondent that "He who knows me only through what I have published does not know me" (Dutens 6.1.65). Du Châtelet also had access to some of Wolff's writings, including a partial translation of Wolff's *Vernünftige Gedanken von Gott, der Welt, und der Seele des Menschens, auch allen Dingen überhaupt* (Janik, 1982, 87), his *Ontologia* (Inst1742, XII, fn.), and his *Eléments d'Astronomie* (Bour and Zinsser, 2009, 273, fn 52; 283, fn 71; 287, fn 79). Here it is worth emphasizing, as we might have emphasized in the case of Leibniz's writings as well, that if there is a single set of doctrines that might be called "Wolff's philosophy"—and to be clear, there is not—it could not be simply read off even from all his writings combined. Any understanding of Wolff's—or Leibniz's—philosophy must necessarily be an interpretation of the many, often obscure, occasionally contradictory, things that they wrote. Derivative interpretations, and even modest synthetic interpretations, can make it sound as if Wolff's or Leibniz's philosophy was simply laying around, ready to be stapled to Newton's physics, or, more ambitiously, to be used as a tool for rethinking the foundations of Newtonian mechanics. But this is fantasy. As ambitious synthetic interpretations are poised to emphasize, if Du Châtelet did have a Wolffian or Leibnizian philosophy, she would have had to construct it from the limited resources available to her. Even if she merely espoused a Wolffian or Leibnizian philosophy, Du Châtelet might nonetheless be reckoned an original thinker in the same sense in which, say, Christian Wolff, Georg Bernhard Bilfinger and Alexander Baumgarten are all regarded by many as both followers of Leibniz and original thinkers in their own right.

The second philosophically ambitious synthetic project would involve Du Châtelet drawing inspiration from a much wider range of thinkers than Leibniz and Wolff. On this point, Du Châtelet's relationship to Descartes's philosophy may serve as an illustrative example (as is underscored, for example, in Detlefsen & Janiak, 2018 and in Detlefsen, 2019). In section V of the Preface to the *Institutions*, entitled "How much we owe to Descartes," Du Châtelet explicitly calls attention to what she sees as the profound influence that Descartes has had on the natural philosophy of her time. She acknowledges that "he was wrong on some points of physics," but nonetheless insists that "human reason owes the most to Descartes" (Inst1742, V). Moreover, she presents "the thinking world" as divided by two systems. But the two systems she has in mind are not the systems of Leibniz and Wolff on the one hand and Newton on the other, but rather the systems of Descartes and Newton (Inst1742, VI). It is hard, we think, to deny that Du Châtelet's philosophical ideas were influenced not only by Leibniz and Wolff, but by Descartes as well. But Descartes is just one of the most obvious of many figures that appear to have played a role Du Châtelet's philosophical reflections. She carefully read Aristotle, Plato, Pope, and Locke.[8] She was well acquainted with the writings of Malebranche and Fontenelle (Janik, 1982, 87). Katherine Brading has convincingly argued that Du Châtelet was profoundly influenced by the works of Newtonian textbook writers such as John Keill, Pieter van Musschenbroek, Willem Jacob's Gravesande, and Henry Pemberton (Brading, 2019, 6). Furthermore, Du Châtelet must have been inspired by her evidently lively discussions with a wide variety of diversely opinionated interlocutors, including Maupertuis, Mairan, Clairaut, Johann Bernoulli II and, of course, Voltaire. Nothing about this is surprising. Asking after the originality of Du Châtelet's philosophy can understandably lead us to focus on individual trees at the expense of the forest. We understandably, even rightly, bear down on specific, narrow questions such as the relationship between Du Châtelet's views on substance and those of Leibniz and Wolff (Stan, 2018). But—as philosophically ambitious synthetic interpretations remind us—it would be a mistake to let our focused attention obscure the big picture. If we locate the originality of Du Châtelet's philosophy in its synthetic nature, we should recognize that the philosophical resources available to her extend far beyond the writings of Leibniz and Wolff that were at her disposal.

[8] For Aristotle, see Inst1742, §7, for Plato see, Philips (1942), for Pope, see Zinsser (2006, 130).

Thus far we have considered derivative and synthetic interpretations of Du Châtelet's philosophy. But a third possibility remains. What we might call *novel interpretations* insist that even philosophically ambitious synthetic views are in danger of underselling Du Châtelet's philosophical originality. Consider, by way of illustration, Leibniz. There is no denying that Leibniz was a deeply synthetic thinker. He was raised on the conciliatory eclecticism of his mentor Jakob Thomasius (for discussion, see Mercer, 2011, 32–39). He drew explicitly on a wide range of traditions, including Platonism (Smith, 2004), Scholasticism (Pelletier, 2014), and Mechanism (McDonough, 2019; Mercer, 2011, 24–48). His texts are almost compulsively riddled with references to an astonishing range of thinkers, literally from Aristotle to Zoroaster. Furthermore, as Patrick Riley has put it, "It was characteristic of Leibniz to try to reconcile apparently conflicting ideas, to take from each kind of thought that which was soundest and to synthesize it with seemingly incommensurable truths of other systems" (1996, 14, see also Dascal, 2008, 1; Mercer, 2011, 23). A philosophically ambitious synthetic interpretation of Leibniz would be right to suggest that he aimed to weave together the views of others to form a novel philosophical outlook of his own. But few would deny that such a reading would nonetheless clearly underestimate the originality of Leibniz's thought. It is true that Leibniz was an ecumenical, elective, synthetic thinker. It is true that there is originality to be found in his attempts to reconcile the views of others. But Leibniz's originality is widely held to go beyond even novel synthesis. From his reflections on theodicy to the labyrinth of the continuum, Leibniz's philosophy is additionally thought to be brimming with ideas and arguments that can only be fairly described as being novel in the strongest possible sense, as being as original to him as any idea ever is to anyone.

Was Du Châtelet a similarly original thinker? Are there genuinely novel philosophical ideas and arguments to be found in her writings? Recent, detailed work has begun to make a strong case for a positive answer to this question (see, for example, Amijee, 2024; Brading, 2019; Carus, 2024; Hagengruber, 2022), and we expect—with enthusiasm—more work to continue in this vein. Here we will content ourselves with noting two considerations that may help to clear ground for future novel interpretations.

Above we cited some modest remarks made by Du Châtelet that might seem to rule out the possibility of novel interpretations. Those remarks - interestingly, written on proof sheets at least partially in another hand (Hagengruber et al., 2024) - included, most strikingly, Du Châtelet's declaration that "[I]n the first few chapters, I'll outline the principal opinions of G. W. Leibniz on Metaphysics, which I have taken from the works of the famous Christian Wolff, whom you have heard me talk so much about with one of his Disciples who stayed with me for some time and who sometimes gave me extracts from his work" (Inst1742, XII). The first consideration is to simply note that Du Châtelet's modest proclamations needn't be taken at face value. Prefatory self-deprecation was a common literary trope in the early modern period. Famously, in his *Epistle to the Reader*, included as a preface to his *Essay Concerning Human Understanding*, Locke acknowledges that his era is blessed with "master-builders, whose mighty designs, in advancing the sciences, will leave lasting monuments to the admiration of society." But Locke modestly does not count himself among those "master-builders," assigning to himself merely the role of "removing some of the rubbish that lies in the way to knowledge" (1998, 11). Descartes' *Meditations*, Hume's *Treatise*, Newton's *Principia*, and even Voltaire's *Candide* are all likewise prefaced with similarly modest or self-effacing remarks. Du Châtelet's deferential comments at the beginning of the *Institutions* no more preclude the possibility of her work containing utterly novel ideas than do similarly modest remarks preclude the possibility of our finding utterly novel ideas in the works of Locke, Descartes, Hume, Newton and Voltaire. Indeed, given the patriarchal expectations of her time,[9] which would have provided her with an almost irresistible reason for at least stated modesty, even less so.

The second consideration is that we have reasons to believe that Du Châtelet—despite her conventional modesty—in fact saw herself as an ambitious seeker of the truth, unconstrained by the views of her predecessors. Her rejection of partisanship is relevant here, and we may quote her admonishment to her son to "Guard yourself … against the inevitable obstinacy to which the spirit of partisanship carries one," as well as her witty declaration that "this [partisan] frame of mind is dangerous on all occasions of life; but it is ridiculous in physics" (Inst1742, VII). Also relevant is her suggestion that "Each philosopher has seen something, and none has seen all," a statement that clearly implies that Du Châtelet thinks that no philosophical system has captured the full truth as she sees it (Inst1742, X). Most striking, however, is a letter of 1740 written by Du Châtelet to Frederick, in which she writes:

> Judge me for my own merits, or lack of them, but do not look upon me as a mere appendage to this great general or that great scholar, this star that shines at the court of France or that famed author. I am in my own right a whole person, responsible to myself alone for all that I am, all that I say, all that I do. It may be that there are metaphysicians and philosophers whose knowledge is greater than mine, although I have not met them. Yet, they are but frail humans, too, and have their faults; so, when I add the sum total of my graces, I confess I am inferior to no one.

This passage, written in private, suggests a very different picture of Du Châtelet's intellectual ambitions, a picture that is almost exactly the inverse of the picture suggested by her self-deprecating remarks in the Preface to the *Institutions*. It is a picture that should encourage proponents of novel interpretations in making their case. For it is a picture that suggests that Du Châtelet at least saw herself as an equal to the great thinkers she cites and engages with, great thinkers such as Newton, Leibniz, and Wolff. Furthermore, it has frequently been noted that in the Preface, Du Châtelet appeals to the metaphor of seeing further by standing on the shoulders of giants—a metaphor that although now strongly associated with Newton, goes back to the Middle Ages. What—as far as we know—has not been noted is that Du Châtelet's version contains a slight, but potentially revealing twist. Although it

[9] See, for example, Frederick's letter to Du Châtelet dated 18 March 1739 (Kölving and Brown, 2018 [Letter 285] 1: 575) and Kant (2011, 37). Even Voltaire, who obviously admired Du Châtelet's intellect, is reported to have written to Frederick that she "was a great man whose only fault was being a woman" (Pagden, 2013, 103).

admitted of many subtle variations, the traditional metaphor has *dwarfs* seeing further by standing on the shoulders of giants (Merton, 1985). As Robert Merton has noted, Newton's version, which reads "If I have seen further it is by standing on ye sholders of Giants," rounds down the traditional metaphor by making it unclear both if Newton is likening himself to a figure of lesser stature and if he thinks that he has in fact seen further (Merton, 1985, 8–9). Du Châtelet's version, which reads "we rise to the knowledge of the truth, like those giants who climbed up to the skies by standing on the *shoulders of one another*," in contrast, rounds up on both points (emphasis added, Inst1742, V). Where the traditional version implies that we are small in comparison with our predecessors, and Newton's version is modestly unclear about his stature and success, Du Châtelet's version is doubly unambiguous. In keeping with the spirit of novel interpretations, Du Châtelet's version suggests that she takes herself to have indeed seen further and reckons herself a no lesser figure than her predecessors, a giant in her own right.

This collection of essays aims to contribute to the important task of understanding the relationship between Du Châtelet's philosophy and the philosophies of Leibniz and Wolff. Was Du Châtelet a devoted follower of Leibniz and Wolff? Or perhaps only of Leibniz? Or perhaps only of Wolff? Alternatively, was Du Châtelet a synthetic, but nonetheless original thinker, and, if so, what exactly did she synthesize and how? Did she primarily reconceptualize Newtonian physics on Leibnizian and Wolffian grounds, or did she weave together a new philosophy of her own? If she did weave together a new philosophy of her own, what were its most important threads? Were they drawn from Leibniz and Wolff or from a wider range of thinkers such as Descartes, Locke, and Newtonian commentators? Or, finally, was Du Châtelet a more thoroughly original thinker than even synthetic interpretations suggest? Would it be more accurate to describe Du Châtelet as using Leibniz and Wolff as springboards for her own ideas and views? Did she stand—as she suggests—like a giant on the shoulders of giants, seeing entirely new things, offering entirely new arguments? The aim of this volume is above all to encourage further attention to the rich, still underexplored, seam where Du Châtelet's philosophy joins with the philosophies of her contemporaries, and especially with the philosophies of Leibniz and Wolff.

The chapters of this volume are arranged roughly in the order of themes addressed in the *Institutions de Physique*, with papers on Du Châtelet's concept of freedom and her moral philosophy following at the end. The volume thus begins with an investigation of the epistemological situation of the human being as set forth by Du Châtelet in the *Institutions de Physique*. In her chapter, *The Certain and the Probable,* Anne-Lise Rey takes up the question posed by Du Châtelet in her main work: "how we can advance our knowledge while at the same time delineating the scope of what we can actually know." Rey investigates Du Châtelet's theory of hypotheses against the backdrop of Leibnizian epistemology. She argues that awareness of the limitations of our faculties, for Du Châtelet, is not a skeptical challenge, but rather a prerequisite for the acquisition of certain knowledge. Rey's chapter aims to make sense of the fact that hypotheses for Du Châtelet are both a response to our epistemic impotence *and* a means to certainty.

The Principle of Sufficient Reason is clearly one of the most important threads common to the metaphysics of Du Châtelet, Wolff, and Leibniz. In her chapter, "Émilie Du Châtelet on Sufficient Reason. A Development in View of Leibniz and Wolff," Clara Carus shows how Du Châtelet develops the concept of *sufficient reason* beyond the views of Leibniz and Wolff. More precisely, she argues that Du Châtelet advances the notion of 'sufficient reason' in two significant ways: First, departing from Leibniz and Wolff, Du Châtelet explicitly delineates the relationship between the distinct concepts of 'cause' and 'sufficient reason'. Second, Du Châtelet implies that a sufficient reason is, by definition, intelligible to human beings. The chapter concludes by showing that the epistemological elaboration of the concept 'sufficient reason' in relation to causes is a step towards invalidating the intelligibility of sufficient reasons in God - an inheritance from Leibniz and Wolff.

Du Châtelet's understanding of extension is both fascinating and puzzling. Many important themes in the *Institutions de Physique*, such as the idea of space, the concept of bodies, and the existence of simple beings hinge on how we understand Du Châtelet's notion of extension. In §77 of the *Institutions de Physique*, Du Châtelet suggests that when "we consider two things to be different, and when we distinguish one from the other, in our minds we place one external to the other." In her chapter *Du Châtelet on External-to and the Idea of Extension*, Qiu Lin enters into a recent debate in the literature concerning whether, in this passage and similar passages, Du Châtelet means to hold that when we represent two different things we must represent them as being spatially or ontologically external to one another. Taking a variety of arguments into account, Lin defends a spatial reading and sets the stage for future work on Du Châtelet's understanding of extension and externality.

Qiu Lin's contribution is nicely complemented by Katherine Dunlop's chapter, *Extension and Space in Chapter V of Institutions de physique*. Dunlop argues that Du Châtelet's chapter on space in the *Institutions de Physique* is an epistemological investigation into the manner in which we represent space and related notions. Dunlop offers a much needed contextualization of Du Châtelet's idea of space in the Early Modern period and considers Du Châtelet's methodological approach to space, as well as the content of the idea of space and of related notions such as extension. The chapter draws comparisons between Du Châtelet's account of space and the accounts offered by Leibniz, Wolff and Kant. Dunlop ultimately argues that Du Châtelet's idea of space and her related idea of extension are not only original but also more convincing than those of her predecessors.

The conception of divine eternity is an important yet puzzling theme of Early Modern philosophy. God is commonly held to be eternal. But in what sense? Does God exist at all times? Is God outside of time? Or is God somehow both in time and outside of time? In his chapter, *Du Châtelet, Wolff, and Leibniz on Divine Eternity*, Jeffrey McDonough distinguishes three views on divine eternity: atemporal, sempiternal, and permanent views. He argues that Du Châtelet is best read as endorsing a permanent view of divine eternity, a view on which God is both in one sense in time and in another sense outside of time. McDonough then compares Du Châtelet's permanent view of divine eternity to the views of Wolff and Leibniz. He argues that while Du Châtelet's permanent view of divine eternity bears some similarities to Leibniz's early views on divine eternity, her position is ultimately closer to Wolff's than to Leibniz's mature understanding of what it means for God to be eternal.

Leibniz famously maintains that "each substance is like a world apart, independent of all other things, except God." In her chapter *Du Châtelet's Rejection of Leibniz's World Apart Doctrine*, Fatema Amijee argues that Du Châtelet rejects Leibniz's World Apart Doctrine. She suggests that Du Châtelet—in radical opposition to Leibniz—insists on the causal connectedness of all created substances. Amijee traces the roots of Du Châtelet's departure from Leibniz's view to two more fundamental anti-Leibnizian commitments, namely, her conviction that natural (i.e., non-divine) sufficient reasons must be accessible to finite minds and her commitment to a causal theory of intentionality.

Du Châtelet, Wolff, and Leibniz all famously—or infamously—hold that extended, continuous, divisible bodies are somehow grounded in extensionless, discrete, indivisible entities. In his chapter, "Infinity for us: Du Châtelet on the Labyrinth of the Continuum," Gabriel Müller focuses on Du Châtelet's understanding of the relationship between extended and unextended entities and weighs objections to that relationship made by Leonard Euler in his *Letters to a German Princess.* Euler sees two main faults with theories of simple unextended beings, namely, that (i) they confuse physical and metaphysical divisibility and (ii) they introduce an incoherent distinction between two types of extension. Müller examines Du Châtelet's answer to Euler's objections and identifies a core disagreement: while extension is an irreducible property for Euler, Du Châtelet holds (with Wolff) that it can be further explained. Müller offers three possible ways of justifying unextended simples and argues that a psychological account is both the most convincing and fits best with Du Châtelet's text.

Du Châtelet wrote an inspiring essay on freedom that was for a long time falsely attributed to Voltaire. In their chapter, "Émilie Du Châtelet on Freedom: The Wolffian Context," Jan Forsman and Jani Hakkarainen help to contextualize Du Châtelet's *On Freedom* by drawing out its relations to Christian Wolff's theory of free will and action. Wolff's account of freedom arose in the context of a dispute with German Pietists concerning the idea of internal spontaneity and its relation to free actions. Forsman and Hakkarainen's show how that context helps to clarify certain puzzling features of *On Freedom*, particularly ones that are concerned with the connection of the will to physical action and the issue of whether Du Châtelet supports a more compatibilist or more libertarian view of freedom. Drawing on competing texts, Forsman and Hakkarainen propose a conciliatory reading and conclude that Du Châtelet is best read as responding to what she calls the "illustrious opponents" of freedom.

The last chapter of the volume is dedicated to a comparison between Du Châtelet and Leibniz regarding the foundations of morality. While much has been written about the relationship between Du Châtelet's and Leibniz's theoretical philosophy, so far, little has been said about possible points of connection between Du Châtelet's and Leibniz's moral ideas. In her chapter, *The Power of Love and the Force of Reason*,

Julia Borcherding argues that the moral philosophies of Du Châtelet and Leibniz bear a number of important similarities as well as significant differences. Borcherding concludes when it comes to the moral realm, Du Châtelet is best seen as effecting a sentimentalist transformation of Leibniz's rationalism. While Du Châtelet holds on to, and even amplifies, certain crucial Leibnizian rationalist principles – including, importantly, the PSR—she nonetheless insists that our passions and desires, though molded and constrained by reason, play a leading role in her understanding of morality.

Collectively, the papers of this volume suggest that the relationships between Du Châtelet's philosophical views and the views of Leibniz and Wolff are incredibly rich and complex, almost certainly too rich be fitted into any one tidy interpretive framework, be it derivative, synthetic, or original. While some of the chapters, like Borcherding's, place Du Châtelet in relatively close relation to Leibniz, other chapters, like Forsman and Hakkarainen's paper, Queryas well as Müller's and McDonough's papers, emphasize similarities to Wolff's views, while also noting important points of departure. Other papers underscore even more strongly the originality of Du Châtelet's thought. Amijee's paper, for example, argues for a clear difference between Du Châtelet's and Leibniz's account of simple substances, while Carus's and Rey's papers argue that although Du Châtelet took Leibniz and Wolff as important sources, she nonetheless pushed them in quite novel directions in the context of her epistemologically grounded metaphysics. Much more, of course, remains to be explored. It is our hope that this volume will encourage others to take up the pressing task of investigating the fascinating points of contact and departure between the philosophies of Du Châtelet, Leibniz of Wolff.

References

Amijee, F. (2024). Du Châtelet's Causal Idealism. *British Journal for the History of Philosophy*, 1–22.

Badinter, E., & Muzerelle, D. (2006). *Madame Du Châtelet: La femme des Lumières*. Bibliothèque nationale de France.

Barber, W. H. (1967). "Mme du Châtelet and Leibnizianism: the genesis of the Institutions de physique", in W. H. Barber, J. H. Brumfitt, R. A. Leigh, R. Shackelton and S. S. B. Taylor (Eds.), The Age of Enlightenment: *Studies Presented to Theodore Besterman*, 200–22. Edinburgh: University Court of the University of St. Andrews.

Besterman, T. (1958). "Les lettres de la Marquise du Châtelet", Theodore Besterman (ed.), Geneva: Institut et Musée Voltaire, 2 volumes.

Brading, K. (2019). *Émilie Du Châtelet and the Foundations of Physical Science*. Routledge.

Carus, C. (2024). Émilie Du Châtelet's Metaphysics in Light of her Concept of 'A Being'. *Journal of Modern Philosophy*, 1–21.

Detlefsen, K. (2019). Du Châtelet and Descartes on the Roles of Hypothesis and Metaphysics in Natural Philosophy. In E. O'Neill & M. Lascano (Eds.), *Feminism and the history of philosophy. The recovery and evaluation of women's philosophical thought* (pp. 97–127). Springer.

Detlefsen, K., & Janiak, A. (2018). Émilie du Châtelet. *Stanford Encyclopedia of Philosophy*. https://plato.stanford.edu/archives/win2018/entries/emilie-du-chatelet/.

Droysen, H. (1909). Die Marquise du Châtelet, Voltaire und der Philosoph Christian Wolff. *Zeitschrift Für Französische Sprache und Literatur, 35*, 226–248.
Efferz, D. (2014). Zur Monade bei Leibniz und Wolff. *Studia Leibnitiana, 46*(1), 64–75.
Ehrman, E. (1986). *Madame Du Chatelet. Scientist, Philosopher and Feminist of the Enlightenment.* Berg.
Frederick II, King of Prussia (1846). *OEuvres de Frédéric le Grand*. Berlin: Imprimerie royale.
Janik, L. G. (1982). Searching for the metaphysics of science. The structure and composition of Madame Du Châtelet's Institutions de physique, 1737–1740. *Studies on Voltaire and the Eighteenth Century, 201*, 85–113.
Hagengruber, R. (2020). Pressure on enlightenment. Criticizing the bible and philosophy. From Nogarola to Du Châtelet. *Dianoia: Rivista di filosofia, 31*, 341–351.
Hagengruber, R. (2022). Du Châtelet and Kant. Claiming the Renewal of Philosophy. In R. Hagengruber (Ed.), *Époque Émilienne* (pp. 75–84). Springer.
Hagengruber, R. (2024). *Teaching women philosophers. Ideas and concepts from women philosophers' writings over 2000 years.* Springer.
Hagengruber, R. E. et al. (2024) AVANT PROPOS. In: Du Châtelet, Émilie: Institutions de physique. The Paris Manuscript BnF Fr. 12265. A Critical and Historical Online Edition, Center for the History of Women Philosophers and Scientists Paderborn, 1.0. https://dcpm.historyofwomenphilosophers.org/documents/view/avant_propos
Hagengruber, R. (N.d.). Émilie Du Châtelet. https://historyofwomenphilosophers.org/project/directory-of-women-philosophers/du-chatelet-emilie-1706-1749/
Hutton, S. (2004). Emilie du Châtelet's Institutions de physique as a document in the history of French Newtonianism. *Studies in History and Philosophy of Science, 35*, 515–531.
Kant, I. (2011). *Observations on the feeling of the beautiful and sublime and other writings.* In P. Frierson & P. Guyer (Eds.). Cambridge University Press.
Kölving, U., & Brown, A. (2018). *La Correspondance d'Émilie Du Châtelet.* Centre International d'Étude Du XVIII E Siècle.
Lascano, M. (2011). Emilie du Châtelet on the existence and nature of god: An examination of her arguments in light of her sources. *British Journal for the History of Philosophy*, 741–758.
Locke, J. (1998). An essay concerning human understanding. In R. Woolhouse (Ed.). Penguin Classics.
McDonough, J. K. (2019). Leibniz's philosophy of physics. *The Stanford Encyclopedia of Philosophy.*
Mercer, C. (2011). *Leibniz's metaphysics: Its origins and development.* Cambridge University Press.
Merton, R. K. (1985). On the Shoulders of Giants: A Shandean Postscript. The vicennial ed. *San Diego: Harcourt Brace Jovanovich.*
Middell, K., & Neumann, H.-P. (2013). *Der Briefwechsel zwischen Christian Wolff und Ernst Christoph von Manteuffel 1738 bis 1748. Transkriptionen aus dem Handschriftenbestand der Universitätsbibliothek Leipzig.* Leipzig: https://slub.qucosa.de/landing-page/%3A%2F%2Fslub.qucosa.de%2Fapi%2Fqucosa%253A2854%2Fmets/%3FL%3D1
Neumann, H.-P. (2022). Émilie Du Châtelet within the correspondence between Christian Wolff and Ernst Christoph of Manteuffel. In R. Hagengruber (Ed.), *Epoche Emilienne* (pp. 147–174). Springer.
Neumann, H.-P., Wells, A., Muller, J., & Hagengruber, R. (N.d.). *The gensis of Emilie Du Châtelet's Institutions de Physique.* https://historyofwomenphilosophers.org/events/
Pagden, A. (2013). *The enlightenment and why it matters.* Random House.
Pelletier, A. (2014). Leibniz and the scholastics. *Studia Leibnitiana*, 123–126.
Philips, E. (1942). Madame du Châtelet, Voltaire and Plato. *Romatic Review, 33*(3), 250–263.
Project Vox Team. (2019). Gabrielle Émilie Le Tonnelier de Breteuil, La Marquise Du Châtelet. https://projectvox.org/du-Châtelet-1706-1749/
Reichenberger, A. (2016). *Émilie du Châtelets Institutions physiques. Über die Rolle von Prinzipien und Hypothesen in der Physik.* Springer.

Seguin, M. S. (2021). Émile Du Châtelet and her Examens de la Bible: A radical clandestine woman philosopher. *Intellectual History Review, 31*(1), 129–141.
Seguin, M. S. (2022a). Early modern biblical criticism and the role of women: The case of Emilie Du Châtelet. In D. Jalobeanu & C. Wolfe (Eds.), *Encyclopedia of early modern philosophy and the sciences* (pp. 523–526). Springer.
Seguin, M. S. (2022b). Madame Du Châtelet, Clandestine Philosopher. In R. Hagengruber (Ed.), *Époche Émilienne* (pp. 441–454). Springer.
Smith, J. (2004). Christian Platonism and the metaphysics of body in Leibniz. *British Journal for the History of Philosophy, 12*(1), 43–59.
Stan, M. (2018). Emilie du Châtelet's Metaphysics of Substance, in: *Journal of the History of Philosophy, 56*(3) 477–496. https://doi.org/10.1353/hph.2018.0048
Wade, I. O. (1941). *Voltaire and Madame du Chatelet—An essay on the intellectual activity at Cirey*. Princeton University Press.
Zinsser, J. (2006). *La dame d'esprit: A biography of the Marquise Du Châtelet*. Viking.
Zinsser, J. (2007). *Emilie Du Chatelet: Daring genius of the enlightenment*. Penguin.
Zinsser, J. P., & Bour, I. (2009). *Émilie Du Châtelet. Selected philosophical and scientific writings*. University of Chicago Press.

Clara Carus is Alexander-von-Humboldt Postdoctoral Fellow at Harvard University. Her research focus is on early modern metaphysics, especially the works of Leibniz, Wolff and Du Châtelet.

Jeffrey K. McDonough is Professor of Philosophy at Harvard University. His primary areas of research lie in early modern philosophy, the philosophy of science, the philosophy of religion, and especially their intersections.

Chapter 2
The Certain and the Probable in Émilie du Châtelet's *Institutions de Physique*

Anne-Lise Rey

The phrase *Hypotheses non fingo* is often regarded as a slogan encapsulating Isaac Newton's methodology. While its generality is often overestimated, it would guarantee certainty in our knowledge of nature. To this old refrain, I would like to contrast a *hypotheses fingo*, an "I imagine hypotheses,"[1] which Émilie du Châtelet uses to think about certainty. She justifies the use of hypotheses in the *Institutions* as both a response to our epistemic impotence and as a means to certainty. The status of hypotheses in the *Institutions* has been well documented,[2] but what sense can we make of our epistemic limitations?

In the preface to the *Institutions*, after reminding us that "truth can take different forms," Émilie du Châtelet points out that all representations in physics are still partial and confused:

> We are still in physics, like this man blind from birth whose sight Chiselden restored. At first this man saw nothing but a blur; it was only by feeling his way and at the end of a considerable time that he began to see well. This time has not quite come for us, and perhaps will never come entirely; there are probably some truths not made to be perceived by the eyes of our mind, just as there are objects, that those of our body will never perceive. But he who refused to learn because of this limitation would resemble a lame person who, having

[1] A.-L. Rey (2022a).

[2] Karen Detlefsen (2019), Sarah Hutton (2012), Ruth Hagengruber, «Émilie du Châtelet between Leibniz and Newton. The Transformation of Metaphysics», in ibid., pp. 1–60; A.-L.Rey (2013).

I would like to thank the journal *Les Etudes philosophiques* where this paper has been originally published in French: «Le certain et le probable dans les *Institutions de Physique* d'Emilie du Châtelet», dossier «Emilie du Châtelet et la certitude» dirigé par A.-L. Rey et Cl. Schwartz, *Les études philosophiques*, Juillet 2023-3, pp. 23–38. Thanks to Ceciel Meiborg for kindly proofreading this text.

A.-L. Rey (✉)
Université Paris Nanterre, IRePh, Paris, France
e-mail: alrey@parisnanterre.fr

C. Carus and J. K. McDonough (eds.), *Émilie Du Châtelet in Relation to Leibniz and Wolff*, Women in the History of Philosophy and Sciences 23,
https://doi.org/10.1007/978-3-031-80541-7_2

a fever, would not take the remedies which can cure it, because these remedies would not stop him from limping.[3]

This means that we need to think about how we can advance our knowledge while at the same time delineating the scope of what we can actually know. Once we hold these two questions together, awareness of our limitations becomes less a sign of a skeptical challenge than the prerequisite for the production of certain knowledge. Émilie du Châtelet clearly indicates that even if certain truths elude us, this should not keep us from our efforts to gain knowledge; our limitations should not lead to renunciation, on the contrary. Put differently, if our cognitive capacities are feverishly lame, our fever can be remedied by clearly distinguishing what our mind can know. Our perception of the sensible world is confused. Since our ideas are always an articulation of obscure representations and clear ideas,[4] our perceptive capacities do not allow us to grasp certain representations very clearly. Fundamentally, however, it is this confusion that allows us to perceive phenomena.

This argument seems to rest on the classical thesis of our epistemic impotence, which in turn is based on the observation of our constitutive imperfection, as we can see in a passage in §134: "[…] as it is impossible for us to represent to ourselves the internal state of all simple Beings (upon which, nevertheless, the Phenomenon of extension depends), all perception of the realities must by our nature escape us."[5] This confusion and obscurity are necessary to guarantee the possibility of each of our points of view on the world, i.e., the possibility of human beings who are able to perceive phenomena. By pointing out the contradiction between clear perception and the very existence of phenomena,[6] du Châtelet highlights the singularity of our epistemic situation: the necessarily confused perception of phenomena is our epistemic identity.

However, if the perception of phenomena is always flawed, is it possible to know anything at all? Either we think that any knowledge is impossible because we cannot know everything, or we claim that in order to know Nature we need a partial point of

[3] Emilie Du Châtelet (2009), preface, VII.

[4] Émilie du Châtelet, *Foundations of Physics*, §132: "That an infinity of dim representations accompany our clearest ideas is something we cannot deny, if we pay a little attention to ourselves." See Leibniz (1969).

[5] Émilie du Châtelet, *Foundations of Physics*, trans. Katherine Brading et al. (2018), available at www.kbrading.org, §134.

[6] Emilie Du Châtelet, Foundations of Physics, trans. Brading et al., §153: "[…] this degree of confusion and imperfection of our organs is necessary for us to see the objects in the way that we perceive them. A Being more perfect than us would have completely different ideas, and would see things completely differently than us; and in order for it to be able to see the same things as we do, and receive the same impressions, it would have to strip itself of the faculty of distinct perception; for the distinction of parts, and the Phenomena that result from their confusion, and that originate from the ensemble, are incompatible. This is why a statue that is made to be placed at a great height seems hideous and rough when we see it up close, and not from the viewpoint for which it was intended, because we form a distinct idea of all the features that together lead to how it looks, and this is so because these features are too large for us to be able to confuse them, up to a certain distance, the point where they have to be to create an agreeable image. […] Thus, a Being with natural prisms for eyes would not have any more idea of white than a deaf person has of sound.".

view. Here we see a gesture that connects epistemology with anthropology, in other words, that indexes our capacity for knowledge in the definition of what it means to be human. Briefly put, Emilie du Châtelet—already—posits the importance of our epistemic situation.[7] Moving away from the conception of our constitutive limitation as a problem toward the delineation of what I can know, she gradually develops a conception of knowledge that takes its limitations and precariousness to be essential to it.

The aim of this article is to highlight the issues that are at stake in the conceptualization of certainty in physics, based on an analysis of the epistemic situation of the knowing subject.

2.1 Uncertainty: Our Epistemic Condition

The epistemic condition that confronts us with our uncertainty is based on the observation of a convergence between the instability of nature and the inaccuracy of our perceptions.[8] Émilie du Châtelet refers to this condition in Chapter II of the Institutions as a "necessary imperfection."[9] She draws two consequences from this: First, certainty can only come from experience, which she demonstrates through the role she reserves for monads and the status she gives to phenomenality. The second consequence is that the mastering of uncertainty by reducing the domain of knowledge to the phenomenal, does not mean that we give up the right to reflect on the existence of an order in nature, but rather questions the status of the search for first elements. Chapter VII, "Of the Elements of Matter," aims to demonstrate the internal coherence of G.W. Leibniz's explanation of matter based on simple beings or monads: The principle of sufficient reason, the principle of indiscernibles, and the principle of continuity are mobilized to assert that simple, indeterminate elements make up the universe.[10] By choosing to answer this classic question with instruments that seemingly stem from the Leibnizian doctrine, Émilie du Châtelet in fact emphasizes the conflict that exists between the imagination and the inability to represent simple beings. This conflict is both a mode of explanation (in a word, extension would not exist if our perceptual acuity were not limited), and a driver of the undermining of

---

[7] Sandra Harding (2012).

[8] See A.-L. Rey, "Hypotheses fingo.".

[9] In Chap. 2, "Of the Existence of God," she writes: "[T]hus, the imperfections in the creature, a succession of his limitations, are necessary imperfections" (Foundations of Physics, trans. Bour and Zinsser, §28). See A.-L. Rey (2022b).

[10] Émilie du Châtelet, Foundations of Physics, trans. Bour and Zinsser, §120: "Now, as sufficient reason obliges us to state that a thing is different from what one is asking about, since otherwise no sufficient reason is provided and the question always remains the same, if one wants to fulfill this principle about the origin of extension, it is necessary to come in the end to something that is without extension, that has no particles, to give a reason for that which is extended and has particles. Now, a being without extension and without particles is a simple being; so, compounds, extended beings, exist because there are simple beings."

the very doctrine that has just been so convincingly presented. The final sentences of §136 indicate that it is impossible to decide whether monads exist or not, and whether they are relevant, but that in any case this question does not affect the certainty of our research into nature.[11] Hence this certainty cannot refer to the primary elements of things, since certainty can only come from the experiments we perform on bodies (and not from the simple elements of which they are composed). It concludes that physical atoms are epistemically sufficient objects[12] that can satisfy our desire for knowledge and in fact the only ones that can be experienced, unlike simple beings. How is this decision to abandon an explanation of the world based on primitive elements that cannot experienced justified?

The order of matter can be understood on two levels: Of course, we can state that matter is composed of simple elements, but since we cannot experience them, the idea that the primitive elements are material or, on the contrary, immaterial, does not fundamentally affect the certainty that we can have in our knowledge of nature. By proposing to distinguish between the constitutive elements of matter, which can be objects of belief or thought but not of knowledge, and objects that I can actually know with certainty because I can experience them, Émilie du Châtelet draws the contours of an epistemic territory: for knowledge to be certain, it must concern objects of the world that we can experience. To what end? To understand the world in which we live, to propose a plausible interpretation of nature. It is perhaps from this starting point that we can recognize what, in the exposition of Leibniz's theory of monads, are indicators of the limits of Émilie du Châtelet's support to for his doctrine.

The limit of her support is to be found on the threshold that distinguishes what we represent through the imagination and can hence be obscure (according to the same Leibnizian doctrine, when the sensible impression stems from something that is far away) and what really is (the theory of monads, which is inaccessible to experience). But this distinction looks very much like a dissonance, or even a divergence. How does Émilie du Châtelet conceive of the relationship between these two explanations of matter? Is it simply a matter of leading us back to a deceptive sensible world? Is it really a matter of conceiving the explanatory principles of nature as inaccessible to the finite, limited human understanding? Nothing is less certain.

Does this mean that we have to abandon all explanatory principles because of their inaccessibility, or accommodate them within the phenomenal field? Chapter VIII, "Of the Nature of Bodies," addresses this question directly. If, as Emilie Du Châtelet says in §153, "this degree of confusion and imperfection of our organs is necessary for us to see the objects in the way that we perceive them,"[13] then the world as we perceive it is the one we have to account for, without invoking any world behind the world, without trying to represent "God's point of view." In "Situated

[11] Émilie du Châtelet, Foundations of Physics, trans. Bour and Zinsser, §136: "but whether they [the monads] are accepted or rejected, our researches on the nature of things will be no less certain; for, in our experiments we never will arrive at these first elements of which bodies are composed and the physical atoms (§172), though in their turn composed of simple beings, are more than sufficient to exercise our desire for knowledge."

[12] See Guillaume Coissard (2021).

[13] Émilie du Châtelet, Foundations of Physics, trans. Brading et al., §153.

Knowledges," Donna Haraway opposes the "god trick" (which amounts to "seeing everything from nowhere"), guarantor of a unified, hegemonic, and totalizing vision of objective knowledge, to the need to think of an "embodied objectivity" (which is literally inscribed in a body, and which does not fundamentally abolish the experience of subjects). This way, she proposes an alternative view of the subjectivity of the knowing subject and of the deceptive and partial character of the perceiving subject. It is thus a matter of "reclaim[ing] the sensory system that has been used to signify a leap out of the marked body and into a conquering gaze from nowhere."[14] A classic objection to this thesis must be faced: In what way can a partial, embedded perspective, in fact, guarantee an objective vision of the world? Beyond the somewhat provocative paradox of this position, the point is to consider the gaze, the point of view, or even the situation as the starting point of an active system of perception. To position oneself is the condition from which one engages in the organization of knowledge; we become responsible by deciding on the demarcations of rational knowledge.

Émilie du Châtelet invites us to consider our epistemic situation as a renunciation of a panoptic position and to explore what it means to know with certainty in the shared world of phenomena. From this it follows that because nature is imprecise, and I have confused and partial perceptions, it is necessary to emphasize our epistemic situation. Does restricting the field of certainty to what we can experience affect our conception of certain knowledge? Chapter IV poses this question in §53:

> The true causes of natural effects and of the phenomena we observe are often so far from the principles on which we can rely and the experiments we can make that one is obliged to be content with probable reasons to explain them. Thus, probabilities are not to be rejected in the sciences, not only because they are often of great practical use, but also because they clear the path that leads to the truth.[15]

It is through the examination of the relationship between certainty and probability that we come to understand what certainty means in physics. A probable reason is an explanation that comes into play when the "true" reason is absent or inaccessible. By forcing us to ask the question whether a probable reason is also still a causal reason, by investing its relation to causality in the name of a rigorous conception of what it means to "explain," Émilie du Châtelet leads us to divide the search for causality into two parts: On the one hand, there are true causes to which we have no access and which therefore, in a (strict) sense, explain nothing. On the other hand, there are plausible substitutes for causality, which, because they result from the application of certain principles (I will come back to this) and can be experienced, properly explain natural phenomena. This could easily lead us back to the conventional division between first and second causes, but the interest of Émilie du Châtelet's presentation here is to show how we overcome this through the question of determination,[16] in that these substitutes integrate indeterminacy as a fundamental dimension of what she defines as knowledge.

---

[14] Donna Haraway (1988).

[15] Émilie du Châtelet, Foundations of Physics, trans. Bour and Zinsser, §53.

[16] Max Kistler (2000): "An explication is scientific to the extent that is not causal."

However, does this mean that we should regard probable knowledge as merely preparatory, as a steppingstone for more certain knowledge to come, or can all knowledge never be anything more than probable to the highest degree?

It would thus be possible to think about uncertainty on its own terms, and to establish the relationship between conjecture and the uncertainty that results from our perceptual capacities. This would enable us to create a unique epistemic framework for thinking in different terms about our knowledge of the natural world. Once it has been established that the convergence between the instability of nature and the imprecision of our perceptions alters the very meaning of the notion of certainty, we need to show the nature of this new epistemic territory, which is one that excludes the possibility of knowing the principles on which our knowledge of nature is based and describes the ways in which certainty can be constituted in the phenomenal realm. My hypothesis is that Émilie du Châtelet, in presenting what she calls Leibnizian epistemology, is actually using it as an argumentative strategy to elaborate and formulate her own epistemology.

In what follows, I begin by recalling how Leibniz conceptualizes certainty in relation to the probable, and then I attempt to show how Émilie du Châtelet works at the very core of this Leibnizian conception to transform the problem.

2.2 The Leibnizian Moment: Where the Certain Meets the Probable

Recall, by metaphysical certainty, Leibniz means, as we know at least from his De modo distinguendi phaenomena realia ab imaginariis of the early 1680s (1683–1686), something of which its opposite would imply a contradiction. As such, metaphysical certainty involves so-called absolute arguments and is therefore demonstrative. Asserting the reality of phenomena, on the other hand, places us in a different register of certainty: moral certainty, which is inferred from the most probable indicators.

In De modo, Leibniz writes:

> We must admit it to be true that the criteria for real phenomena thus far offered, even when taken together, are not demonstrative, even though they have the greatest probability; or to speak popularly, that they provide a moral certainty but do not establish a metaphysical certainty, so that to affirm the contrary would involve a contradiction.[17]

But what is non-metaphysical certainty? In the margins of his Introduction to a Secret Encyclopedia,[18] G.W. Leibniz distinguishes logical certainty from moral certainty, which he in turn distinguishes from physical certainty. By moral certainty, G.W. Leibniz means "everything that is confirmed by numerous clues that can hardly coincide if it is not true." He makes this very clear in a letter to Simon Foucher in

---

[17] A VI, 4B, 1500-04, translated as G.W. Leibniz, "On the Method of Distinguishing Real from Imaginary Phenomena," in Philosophical Papers and Letters, 364.

[18] G.W. Leibniz (1973).

1675: "The connection of all appearances with each other, which dreams lack, is an a posteriori sign of a real being; but it is not a demonstrative sign."[19] The a posteriori sign here is opposed to demonstration, but how can we have a non-demonstrative access to the real, and what is it worth? To answer these questions, we need to understand, on the one hand, the convergence of evidence as a criterion of truth, to use Leibniz's terms, and, on the other hand, the correlation between this convergence and the degree of probability it seems to induce.

Alongside this moral certainty, but distinct from it, G.W. Leibniz, again in the Introduction to a Secret Encyclopedia, characterizes physical certainty as the convergence of experience: "everything which men have experienced always and in many ways," and he gives an example: that "iron sinks in water," an example that refers to the physical law according to which the density of iron will always be greater than that of water, more precisely than the pressure that water exerts on the object.

This invites us to distinguish the kind, and perhaps even the quality, of certainty and to raise the question of their possible hierarchy: Is moral certainty, for example, less "reliable" than metaphysical certainty? Or is moral certainty simply of a different order than metaphysical certainty?

This also leads us to pose the question of the relationship between the objects we seek to know and the kind of certainty to which we have access. Briefly put, are the objects of physics, such as the laws of motion, susceptible only to physical certainty, characterized as certainty resulting from a consensus or convergence of multiple experiences—Leibniz's text speaks here of "conspiracy"—or can they be grasped and if so, under what conditions, with logical certainty, the first principle of a posteriori thought?

This raises at least two questions: First, what kind of relationship to experience does this category of non-metaphysical certainty enable us to think? In what way does experience intervene in the constitution of these two forms of certainty, physical and moral: can the convergence of experiences be related to the consensus of indices, and can we thus understand this congruence as a particular example of an index? And if so, how does experience relate to the non-demonstrative proof that is established here? A second question is: What is the relationship (or what are the relationships) between certainty and probability? As we have seen, moral certainty is not demonstrative; it constructs its relationship to truth by summoning degrees of probability (as the definition in the Introduction to a Secret Encyclopedia makes clear). But probability itself, as Leibniz asserts in a letter to Hermann Conring of January 1678, can be demonstrated:

> Even in the case of facts, when both sides argue with suppositions and conjectures, it is possible to define exactly, from the given circumstances, which side has the greater probability. In this way, probability itself can be demonstrated and its degrees can be estimated.[20]

This implies that qualitative elements are included in the evaluation of the strength of a reasoning. The relationship between demonstration and moral certainty is thus

[19] See A VI/4-A, 307, 15–17.

[20] G. W. Leibniz, Die Philosophische Schriften, ed. K. I. Gerhardt, 1860, I, 187.

twofold: the moral certainty is not the result of a demonstration, but it does use a demonstrative logic of probability in order to assert itself as true.

We must therefore resolve the problem of moral certainty at the point of its encounter with the probable, that is, at the very point where the absence of demonstration specific to moral certainty accommodates, in compensatory manner, as one might consider it, a logic of the probable.

The Recherches générales sur l'analyse des notions et des vérités (General Inquiries on the Analysis of Notions and Truths) posits three strong theses that sum up Leibniz's position on this point. In a contingent truth, strictly speaking, we never arrive at a demonstration, as Leibniz defines it as a reduction to the identical. What form of proof can we then access, given that Leibniz asserts in §62 that "every true proposition can be proved," i.e. that it can be accounted for by resolution? On what basis does Leibniz speak of contingent truth? The problem is well known: if truths of fact require infinite regress for their resolution, and if only God knows contingent truths since perfect demonstration surpasses any finite intellect (§131), how can we know contingent truths? According to the conventional view that our finite understanding is always taken to be lacking and radically different from God's infinite understanding, which is always thought in relation to the power of this divine understanding, this question returns us to our status as limited epistemic subjects. But we could pose the problem differently, that is, from another perspective, by assuming our epistemic situation: What means do we have at our disposal to know the contingent, and to what form of certainty do we then have access?

To the question of how we can prove the truth of a contingent proposition, G.W. Leibniz offers a first answer in §134 of the Inquiries: reminding us that, by definition, one might even say by nature, contingent truths cannot be completely resolved ("reduced to the identical"), nevertheless, "if the resolution is continued further and further, it continuously approaches the identical propositions, but never reaches them."[21] Only God has certainty about all contingent truths. Leibniz then introduces a comparison: "[...] the difference between necessary and contingent truths is the same as that between intersecting and asymptotic lines or commensurable and incommensurable numbers."[22] Geometry and arithmetic are thus the domains in which Leibniz constructs an analogy that allows us to visualize our access to certainty and to understand, through comparison, the limits of the kind of certainty that we can access here.[23] This is the heuristic function of his analogy.

---

[21] G.W. Leibniz, Inquiries, §134, p. 111.

[22] Ibid., §135, p. 111.

[23] Id., « Vérités nécessaires et vérités contingentes », in ibid., p. 340 : « Dans une vérité contingente, bien que le prédicat soit véritablement dans le sujet, on ne parvient pourtant jamais à la démonstration, c'est-à-dire à l'identité, quand bien même la résolution des deux termes serait indéfiniment poursuivie; il n'appartient qu'à Dieu seul, qui comprend par un seul acte tout l'infini, de voir complètement comment un terme est dans un autre et de comprendre a priori la raison parfaite de la contingente. Pour les créatures l'expérience a posteriori y supplée. C'est pourquoi les vérités contingentes sont en quelque façon aux vérités nécessaires comme les raisons sourdes ou celles des nombres irrationnels aux raisons énonçables des nombres rationnels. »

There are two things we can learn from this: first, there is the problem that if God alone, whose mind encompasses the infinite, can know "all contingent truths with certainty,"[24] does this mean that we can never have certainty in the domain of contingent things? Or is it simply to say that this certainty can never encompass all contingent truths?

But if we are not convinced by this continual extension of asymptotes and we stop taking it as a simple model of intelligibility, as a mere analogy, and instead take it as a proof of the truth of contingent things, Leibniz sets out the two ways of knowing contingent truths: reason and experience. In De Libertate (On Freedom), he states that we know "by experience when we perceive a thing sufficiently distinctly through the senses."[25] A contingent truth we prove through reason "by showing that there is more reason for that which has been done than there is for its opposite."[26]

I would like to start from this indication, question its meaning and the relationship between these two ways of knowing contingent truths. Is it a *connubium* or are they two different ways of access without any necessary correlation or compatibility? How can we determine what it means to "perceive a thing sufficiently distinctly through the senses"?

The question of the status of experience in Leibniz's theory of knowledge could, at first glance, be interpreted as a convenient means of illustrating a theory, of producing an "image of reality." This would be limited to confused knowledge[27] in the sense of Leibniz's Meditations on Knowledge, Truth, and Ideas, i.e., incomplete or approximate knowledge that comes from the testimony of the senses:

> [...] when I cannot enumerate one by one the marks which are sufficient to distinguish the thing from others, even though the thing may in truth have such marks and constituents into which its concept can be resolved. Thus we know colors, odors, flavors, and other particular objects of the senses clearly enough and discern them from each other but only by the simple evidence of the senses and not by marks that can be expressed.[28]

What follows offers the possibility of building a perceptive community that can replace the rational community:

> So we cannot explain to a blind man what red is, nor can we explain such a quality to others except by bringing them into the presence of the thing and making them see, smell, or taste it, or at least by reminding them of some similar perception they have had in the past.[29]

A "sufficient perception" would therefore be that which makes a world[30] in the sense that it provides the means to identify a common world that we all perceive, albeit from different points of view, and which we perceive as common

---

[24] G.W. Leibniz, Inquiries, §134, p. 111.

[25] G.W. Leibniz (1989).

[26] Ibid., The Source of Contingent Truths, in Philosophical Essays, p. 101.

[27] Ch. Leduc (2009).

[28] G.W. Leibniz, Meditations, 291.

[29] Ibid.

[30] Michel Fichant (1998).

because it presupposes a perceptual correspondence between the different convergent particularizations of a global representation of the world.

In this sense, we are already beginning to answer a second problem: How can we determine "that there is more reason in favor of what is happening"? What does it mean to have "more reason"? G.W. Leibniz indicates in De Contingentia[31] that necessity ex hypothesi must be clarified by distinguishing "the necessity of the consequence" from "the necessity of the consequent":

> For, if by definition a necessary proposition is one whose truth can be demonstrated with geometrical rigor, then indeed it could be the case that this proposition is demonstrable: "every truth and only a truth that has greater reason," or this: "God always acts with the highest wisdom." But from this one cannot demonstrate the proposition: "contingent proposition A has greater reason [for being true]," or "contingent proposition A is in conformity with divine wisdom." And therefore it does not follow that contingent proposition A is necessary. So, although one can concede that it is necessary for God to choose the best, or that the best is necessary, it does not follow that what is chosen is necessary, since there is no demonstration that it is the best. And here the distinction between the necessity of the consequence [necessitas consequentiae] and the necessity of the consequent [necessitas consequentis] is in some way relevant; in the end, the proposition in question is a necessity of the consequence, not of the consequent, because it is necessary once we grant the hypothesis that we take it to be the best, assuming that the best is necessarily chosen.[32]

Necessity ex hypothesi is a necessity that is reconstructed a posteriori, on the basis of what has happened, considered as the best: it is a reconstruction after the fact. The mere fact that the choice for the best is necessary does not mean that the contingent proposition that is derived from it is itself necessary. This raises the question of its articulation: How to think about the relationship between this sufficient perception, which presupposes the identification of a perceptive community, and this necessity *ex hypothesi*, which reconstructs an order after the fact?

Achieving true knowledge through experience and not only by doing experiments, presupposes that we perceive the existence of a common world through the connection of phenomena, that is, a world endowed with an order that is revealed after the fact and hence allowing us to reconstruct a coherence. It is therefore these muted propositions, or contingent propositions, that need to be thought, and more precisely, the contingent propositions that we find in the domain of the natural sciences and that result from experiments that are conducted in order to circumscribe the epistemic use of experiments.

Émilie du Châtelet did not have access to all of the texts I have drawn on in the preceding. This discussion is serves more to clarify Leibniz's position than to identify exactly on which sources she bases her thinking. Nevertheless, we can retain three crucial features with which Émilie du Châtelet works to develop her epistemology: the demonstration of the probable, the relationship between certainty and the order of nature, and the status of the phenomenal field. This means taking Leibnizian epistemology, as presented by Émilie du Châtelet, as a field of inquiry that allows problems to become apparent and offers conceptual tools.

---

[31] G.W. Leibniz, On Contingency, in Philosophical Essays, 28–30.

[32] Ibid., p. 30.

2.3 Reducing Uncertainty or How to Render Knowledge Certain?

The Institutions de Physique reflects on the community of perception, questions the irreducibility of the distinction between contingent and necessary truths, and considers the demonstrability of moral certainty. Émilie du Châtelet uses her presentation of G.W. Leibniz's doctrinal position to elaborate and present her own philosophy using a dual strategy.

The first strategy is to identify the principle with the axiom in §3 and §4 of the first chapter and to blur the irreducibility of the oppositions between the two types of truth (necessary and contingent). This leads us to question the transformation of the meaning of metaphysics in this text. This is crucial for the understanding of the meaning of the well-known formula "physics cannot do without the metaphysics on which it is based."[33] In §§2–3, she moves from the uncertainty so prevalent in science to certainty through her choice of principles.

Émilie du Châtelet's double achievement here is that she introduces the principle of non-contradiction into the heart of geometric demonstration by referring to Euclid, and thus demonstrates the logic of the principle without giving it an overarching metaphysical status. By invoking the Pyrrhonians, she shows that this principle is used regardless of the school of thought to which one belongs, including a skeptical school, which rejects the possibility of establishing doctrines.

In what follows in this first chapter (§§4–5), she presents the principle of contradiction as an axiom[34] that is the foundation of truths and certainties. How are we to understand the principle as an axiom? The principle of contradiction makes it possible to define the possible and the impossible. She aims here to counter the Cartesian position that holds that the possible and the conceivable produce a clear idea by pointing out that the conceivable must also be non-contradictory.

Émilie du Châtelet makes two points here. First, a clear idea may turn out to be false; clarity is therefore not enough to decide the truth of an idea. Second, an idea may seem obvious when taken individually, but when it is articulated together with the other ideas that make up the whole description or definition of a reality, it may turn out to be false or absurd. What she points out here is the inadequacy of clarity as a criterion for the truth of an idea. But she also delimits the scope of the principle of contradiction to that of necessary truths, which are already determined and determinable in a single way, e.g., geometrical truths. Émilie du Châtelet distinguishes these truths from another category of truths, i.e., contingent truths. The latter category of truths

---

[33] Lettter 286, Émilie du Châtelet to Frederick the Great, avril 1740, in La correspondance d'Émilie du Châtelet, ed. Ulla Kölving and Andrew Brown (Ferney-Voltaire: Centre international d'étude du XVIIIe siècle, 2018), 576.

[34] Émilie du Châtelet, Foundations of Physics, trans. Bour and Zinsser, §4: "*Contradiction* is that which simultaneously affirms and denies the same thing; this principle is the first axiom, on which all truths are founded. Everyone readily agrees on this, and it would even be impossible to deny it without lying to one's conscience [...]. This axiom is the foundation of all certainty in human knowledge. For, if one once granted that something may exist and not exist at the same time, there would no longer be any truth [...]."

presupposes another principle in order to be established, the introduction of which is justified by the fact that in this other area of validity, that of contingent truths, a principle is needed that is adapted to the fact "that a thing can exist in different ways" (recall the preface, "truth can take different forms"). Here lies the reason for the introduction of a second Leibnizian principle, the principle of sufficient reason: Nothing happens without a reason, and we must go back up the chain of reasons to identify the reason of an action.[35]

Another remarkable dimension is the singularity of Émilie du Châtelet's appropriation of the principle of sufficient reason. While for G.W. Leibniz it seems impossible to return to the reason that grounds the chain of determinations, in her model we proceed until we reach the reason for action, and thus take a step back from the Leibnizian model toward one that supports this very human use of the principle.

In paragraph 8, this principle becomes the condition for the endurance of a discourse of truth, because without the principle of sufficient reason, "truths, for us, would only exist for an instant."[36] She illustrates this with the example of a room left in order, but soon discovered in disarray without anyone having entered. In this way, she opens up a space for conceptual elaboration that allows us to imagine a world without sufficient reason, that is, without God, and therefore a random world. This is not to say that the hypothesis of a world without sufficient reason is correct or that it is endorsed by Émilie du Châtelet. Rather, we should be interested in what she emphasizes in her description of this rejection, namely the correlation between the principle of reason and the stability and permanence of nature.

The second strategy is revealed in her attempt to "bring it [metaphysics] closer to [us]," as the preface states.[37] In doing so, she emphasizes the need to rethink the meaning of metaphysics. This raises the question of whether, even in this 1740 text, she participates in the movement to redefine metaphysics, which was presented par excellence a few years later by Jean Le Rond D'Alembert, for example in the Encyclopédie[38] entry "Éléments des sciences" (Elements of the sciences). In this text, Jean Le Rond D'Alembert distinguishes two meanings of metaphysics, the second of which we are interested in here, which considers that there is a metaphysics specific to each science, understood as the general consequences of our observations.[39] The

---

[35] See Émilie du Châtelet, Foundations of Physics, trans. Bour and Zinsser, §11.

[36] Émilie du Châtelet, Foundations of Physics, trans. Bour and Zinsser, §8.

[37] Émilie du Châtelet, Foundations of Physics, trans. Bour and Zinsser, preface, XII.

[38] See http://enccre.academie-sciences.fr/encyclopedie/article/v5-851-0/.

[39] "Here I am taking Metaphysics in the most rigorous sense, since it is the science of purely spiritual beings. What I am saying here will be even truer when we look at Metaphysics in a more general sense as that universal science which contains the principles of all the others. For if each science is based on observation and on observation alone for its true principles, the Metaphysics of each science can consist only of general consequences resulting from observation and presented from the widest vantage point possible" (Jean-Baptiste le Rond d'Alembert, art. "Elements of the sciences," trans. Lauren Yoder, The Encyclopedia of Diderot & d'Alembert Collaborative Translation Project [Ann Arbor: Michigan Publishing, University of Michigan Library, 2011]: http://hdl.handle.net/2027/spo.did2222.0001.133 [accessed June 9, 2024]. Originally published as "Elémens des sciences," Encyclopédie ou Dictionnaire raisonné des sciences, des arts et des métiers, 5:492 (Paris, 1755).

reconfiguration of the meaning of metaphysics has consequences for delineating a boundary for the construction of certain knowledge in the phenomenal field. In this way, Émilie du Châtelet justifies the delimiting of a phenomenal field within which we can know with certainty is justified in this way. This involves the claim that "the reason of effects" does not require access to the primary reason of things: phenomena can explain phenomena. But does this imply taking up the idea of a perceptive community mentioned in Part 2 and giving it a different role? If this is indeed the case, then the perceptive community guarantees certain knowledge of phenomena, situating the reason for these phenomena at the heart of the phenomenal world.

But in order to achieve this, we must ask ourselves how we can go about a demonstration of the probable. Chapter IV of the Institutions sets out to answer this question: In §57 it says that certainty is "not far from demonstration," in §58 it says that probability is "almost equivalent to a demonstration," and in §67 it goes on to say that "as a very great degree of probability gains our assent, and has on us almost the same effect as certainty, hypotheses finally become truths when their probability increases to such a point that one can morally present them as certainty."[40] We can thus understand how Émilie du Châtelet reconfigures G.W. Leibniz's epistemology by asserting that the phenomenal realm – whose precariousness must be pointed out – is susceptible to an order that we can explain, provided that we introduce a reflection on the probable that makes up its core. From then on, the above-mentioned approximation ("almost," "not far," "morally present them as") is less the mark of the shortcoming of this knowledge than the sign of a pervasive precariousness that is crucial for understanding the new conceptualization of knowledge in the mid-eighteenth century. When Émilie du Châtelet writes that the "degree of probability gains our assent," so that hypotheses become truths (§67), or that we can "morally present as certainty," she indicates the status of this knowledge of nature: a precarious knowledge that accounts for the very instability of the world. But it also indicates that this "analog of a rational deductive system,"[41] as François Duchesneau writes of G.W. Leibniz, is no longer to be thought of, in the Institutions, as a meagre substitute in the absence of a rationally founded system; it is rather the only possible access to knowledge of nature. As such, it profoundly changes the framework of the intelligibility of nature, and thus what it means to know: from now on, knowing implies the integration of approximation as one of its fundamental dimensions.

2.4 Conclusion

In §53, Émilie du Châtelet, in a passage already mentioned, praises probabilities for "clear[ing] the path that leads to truth." While probabilities may have been seen as a convenient steppingstone on the way to the truth, I have attempted to show that, taken

[40] Émilie du Châtelet, Foundations of Physics, trans. Bour and Zinsser, §§57, 58, 67.

[41] François Duchesneau (1993).

to their acme, they are in fact the new face of truth within the framework of knowledge that Émilie du Châtelet helps to forge. To do this, she had to take the Leibnizian epistemology to heart and articulate the idea of a perceptive community that leads us to identify an order within the phenomenal field, which means that producing a demonstration of the probable is possible. On this condition, our epistemic situation can produce a morally certain knowledge of nature.

References

Coissard, G. (2021). Du Châtelet entre monadisme et atomisme: la matière dans les Institutions de physique. *Revue D'histoire des Sciences, 74*(2), 297–329.

Detlefsen, K. (2019). Du Châtelet and Descartes on the Roles of Hypothesis and Metaphysics in Natural Philosophy. In E. O'Neill & M. P. Lascano (Eds.), *Feminist history of philosophy: The recovery and evaluation of women's philosophical thought* (pp. 97–127). Springer.

Duchesneau, F. (1993). *Leibniz et la méthode de la science* (pp. 208–209). PUF.

Fichant, M. (1998). De l'individuation à l'individualité universelle (pp. 143–62). In Science et métaphysique dans Descartes et Leibniz. PUF.

Du Châtelet, É. (2009). *Foundations of physics*. In Zinsser, J. P. (Ed.), *Selected philosophical and scientific writings* (trans. I. Bour & J. P. Zinsser). The University of Chicago Press.

Haraway, D. (1988). Situated knowledges: The science question in feminism and the privilege of partial perspective. *Feminist Studies, 14*(3), 575–599, 581.

Harding, S. (2012). Rethinking standpoint epistemology: What is 'strong objectivity'? *The Centennial Review, 36*(3), 437–470.

Hutton, S. (2012). Between Leibniz and Newton: Émilie du Châtelet and Samuel Clarke. In R. Hagengruber (Ed.), *Émilie du Châtelet: Between Leibniz and Newton* (pp. 77–96). Springer.

Kistler, M. (2000). *Causalité et lois de la nature* (p. 9). Vrin.

Leduc, C. (2009). Substance, individu et connaissance chez Leibniz (p. 192 sq). Vrin & Presses de l'université de Montréal.

Leibniz, G. W. (1969). Meditations on knowledge, truth, and ideas. In L. E. Loemker (Ed.), *Philosophical papers and letters* (pp. 291–295). Reidle.

Leibniz, G. W. (1973). Introduction to a secret encyclopedia. In G. H. R. Parkinson (Ed.), *Philosophical writings* (p. 9) (trans. M. Morris). J. M. Dent & Sons.

Leibniz, G. W. (1989). On freedom. In *Philosophical essays* (p. 96) (trans. R. Ariew & D. Garber). Hackett.

Rey, A.-L. (2013). Le leibnizo-newtonianisme: La construction d'une philosophie naturelle complexe dans la première moitié du 18e siècle. La méthode d'Émilie du Châtelet entre hypothèses et experiences. *Dix-huitième siècle, 45*, 115–129 (2013).

Rey, A.-L. (2022a). L'expérience épistémique comme dialogue entre les Institutions de physique et les Examens de la Bible. *Émilie Du Châtelet Et La Littérature Philosophique Clandestine, Special Issue of La Lettre Clandestine, 30*(2022), 17–39.

Rey, A.-L. (2022). Hypotheses fingo: Certitude et imagination dans les Institutions de physique d'Émilie du Châtelet. In U. Kölving & A. Brown (Eds.), *Émilie du Châtelet, son monde, ses travaux* (pp. 359–371). Centre international d'étude du XVIIIe siècle.

Anne-Lise Rey is Professor of Philosophy at Paris Nanterre University. Her research focuses on the Early Modern Period.

Chapter 3
Émilie Du Châtelet on Sufficient Reason: A Development in View of Leibniz and Wolff

Clara Carus

Abbreviations

A	Gottfried Wilhelm Leibniz: Sämtliche Schriften und Briefe, ed. Deutsche Akademie der Wissenschaften (Darmstadt and Berlin: Akademie-Verlag, 1923–).
AG	Gottfried Wilhelm Leibniz: Leibniz: Philosophical Essays, ed. and trans. Roger Ariew and Daniel Garber (Indianapolis: Hackett, 1989).
DM	Gottfried Wilhelm Leibniz: Discourse on Metaphysics, ed. and trans. Gonzalo Rodriguez-Pereyra (Oxford: Oxford University Press, 2020).
G	Gottfried Wilhelm Leibniz: Die Philosophische Schriften von Gottfried Wilhelm Leibniz, 7 vols., ed. C. I. Gerhardt (Berlin: Weidmann, 1875–90; reprint ed. Hildesheim: Olms, 1960).
GM	Christian Wolff: Vernünftige Gedanken von Gott der Welt und der Seele des Menschen, Auch allen Dingen überhaupt, Halle 1720, Digitalisierung von Drucken aus dem 18. Jhd., Universität und Landesbibliothek Sachsen-Anhalt.
Inst1742	Émilie Du Châtelet: Institutions physiques de madame la marquise Du Chastellet adressés à M. son fils: Nouvelle édition, corrigée et augmentée considérablement par l'auteur (Amsterdam: Aux dépens de la Compagnie, 1742). Translations from passages that are identical in the 1740 edition mostly follow Zinnser/Bour and Brading et. all; the other translations are my own.
Inst1740eB	Émilie Du Châtelet: Institutions de Physique (Paris: Prault, 1740). Work in progress translation by Katherine Brading et al. available here: https://www.kbrading.org/translations

C. Carus (✉)
Department of Philosophy, Harvard University, Cambridge, MA, USA
e-mail: claracarus@fas.harvard.edu

C. Carus and J. K. McDonough (eds.), *Émilie Du Châtelet in Relation to Leibniz and Wolff*, Women in the History of Philosophy and Sciences 23,
https://doi.org/10.1007/978-3-031-80541-7_3

NE	Gottfried Wilhelm Leibniz: New Essays on Human Understanding, ed. and trans. Peter Remnant and Jonathan Bennett (Cambridge: Cambridge University Press, 1981).
Ont	Christian Wolff: Philosophia Prima, sive Ontologia, Gesammelte Werke. (New York: G. Olms, 1980-).
PM	Du Châtelet, Émilie: Traité de metaphisique. The Saint Petersburg Manuscripts. A Critical and Historical Online Edition (2020-2023). Edited by Ruth E. Hagengruber, Andrew Brown, Stefanie Ertz, Ulla Kölving. https://stp.historyofwomenphilosophers.org/documents/view/traite_de_metaphisique#head7

3.1 Introduction

Scholarship on Du Châtelet has developed significantly in the recent decade. However, some crucial aspects of her metaphysical and epistemological framework have not been studied in detail; among them her understanding of the concept 'sufficient reason'. Despite the concept 'sufficient reason' being of central importance to the *Institutions Physiques* there is not a single study dedicated to her understanding thereof. An investigation into her understanding of sufficient reason is all the more pressing considering the fact that Du Châtelet's elaboration of the principle of sufficient reason is a highly original and significant development out of and away from Leibniz and Wolff.

Du Châtelet, departing from Leibniz and Wolff, elaborates on the concept of sufficient reason in two significant ways:

Firstly, Du Châtelet delineates the relationship between the two distinct concepts 'cause' and 'sufficient reason'. Leibniz indicates in various passages that 'sufficient reason' and 'cause' are not the same, yet he continues to use the terms interchangeably throughout his oeuvre and does not clarify conceptually wherein the distinction between cause and sufficient reason lies, as Wolff also points out (Ont, §71). Wolff makes explicit that the terms sufficient reason and cause do not signify the same thing, however his focus lies solely on the concept of sufficient reason. Put briefly, Wolff considers the concept 'cause' unworthy of a theoretical investigation in itself. In turn, the elucidation of what it is that constitutes a cause in distinction from a sufficient reason and especially the relation between the two is lacking in Wolff. Du Châtelet, as we will see, draws a clear distinction between the two concepts and spells out how 'cause' and 'sufficient reason' relate to each other. This is indispensable as it shows that sufficient reason signifies a relation to a cause that is intelligible and roots out the error of positing a sufficient reason that cannot be known.

Secondly, Du Châtelet puts forward an epistemological theory of the knowledge of beings in their essential, attributive and modal determinations. This theory has important roots in Wolff.[1] However, there are differences to Wolff in Du Châtelet's account of how a sufficient reason determines the cognition of beings, which do not simply go back to Leibniz either. The most important difference lies in the fact that for Du Châtelet essences are solely intelligible in themselves (on the basis of the principle of contradiction (henceforth PC)) and not by manner of the way they come about, i.e. through a cause or reason. Du Châtelet holds that attributes and modes of beings always have a sufficient reason, while the essences of beings do not have a further reason: Why beings are the way they are in their essence, i.e. why beings in terms of creation are the way they are, is not philosophically answerable. In this vein, Du Châtelet develops a clearer and epistemological account of sufficient reason and its role in our knowledge of beings.

Both these aspects in which Du Châtelet develops the understanding of the intelligibility of a sufficient reason further are fundamental developments in the epistemology of the Early Modern Period and call for further investigation. This paper provides a first overview over the epistemological development of sufficient reason in Du Châtelet. Her account exacts the pressing question: is a sufficient reason that is not intelligible to the human being a conceptual contradiction or, to speak in Du Châtelet's terms, a chimera?

3.2 The Distinction and Relation Between Cause and Sufficient Reason

In Du Châtelet, the understanding of what constitutes a sufficient reason was greatly advanced by clearly distinguishing the concept 'sufficient reason' from the concept 'cause', as well as by elucidating the relationship between the two. Thus, by examining the distinction between cause and sufficient reason, as well as their relationship, we will be in a position to determine in what sense Du Châtelet further developed the concept of sufficient reason in view of her sources, Leibniz and Wolff respectively.

In his formulations of the principle of sufficient reason (hereafter PSR), Leibniz sometimes uses the term cause [causa], sometimes the term reason [ratio], and sometimes both at the same time, as for example in his essay *Specimen inventorum de admirandis naturae arcanis,* where he writes:

> There are two principles of all reasonings, the principle of contradiction and [...] the principle that a reason must be given, i.e. that every true proposition, which is not known per se, has an a priori proof, or that a reason can be given for every truth, or, as is commonly said, that nothing happens without a cause.[2] (G VII, 309)

---

[1] Wolff's theory of how attributes and modes are to be understood on the basis of 'sufficient reason' has important roots in Leibniz's theory of predication. Leibniz claims that predicates always have an intelligible 'sufficient reason' in the subject or in the nature of the subject. However, for Leibniz it is the case that this intelligibility is in its full sense reserved for God (cp. DM; 7).

[2] Cp. also New Essays A VI, 6, 475. Leibniz does, however, distinguish between different concepts of cause. For example, Bobro (2021) points to several passages where Leibniz employs 'causa efficiens', 'first cause', or first truth as a cause. Furthermore, Leibniz uses 'first cause', 'efficient

Leibniz often identifies 'reason' with 'cause', as he does in this passage, or uses the words 'cause' and 'reason' interchangeably (e.g. CP, 101; A VI, 6, 475; for further reading see Pikkert, 2018). There are, however, some rare passages in which Leibniz distinguishes generally between 'cause' and 'reason'. In *The Elements of True Piety*, for example, he writes:

> [T]here is nothing without a reason [ratione], but that does not mean that there is nothing without a cause [causa]. For a cause is the reason for a thing outside of the thing, or its reason of production, but it is possible that the reason for a thing is inside the thing itself. And this is the case in all those things which are necessary, like the truths of mathematics which contain their reason in themselves; likewise God, who alone is the actual reason for the existence of actual things. (A VI 4, 1360)

Hence, the reader can develop a notion of what it is that distinguishes a 'sufficient reason' from a 'cause' on the basis of Leibniz's examples and his different use of the terms, but there is no stringent conceptual distinction drawn between cause and sufficient reason in Leibniz's works and in connection with this he does not establish a necessary relation between sufficient reason and the intelligibility of the cause.

Wolff, in his work on the concept of sufficient reason, explicitly builds on Leibniz and writes that Leibniz has, in his use of the PSR, "illuminated the way for him" and enabled him to develop a distinct concept [notionem distinctam] of sufficient reason (Ont, §71). Wolff praises Leibniz for possessing, unlike the scholastics, a clear concept of sufficient reason but criticizes him for not elucidating the difference between cause and reason, as stated above.[3] Wolff defines his concept of sufficient reason as follows: "By *sufficient reason* we understand that through which it is understood why something is"[4] (Ont, §56). However, it is surprising that although Wolff explicitly wishes to distinguish between the concepts 'reason' and 'cause' in Ont, §71, and refers to his definition of sufficient reason in §56 in this context, he does not define the concept of cause more closely. We can assume that Wolff considered the definition of cause to be clear or not worthy of further investigation. We can see this problem in Wolff more clearly in the example he gives in the *German Metaphysics*:

> If a thing A contains something in itself through which one can comprehend why B is, (B may be something in A or outside of A) then we call what is found in A the reason of B; A itself is called the cause and of B we say that it is grounded in A. I wish to elucidate this through an example. If I investigate how it transpires that everything grows quickly in a garden, and find that it is to be attributed to the warmth of the air, then the warmth is the reason [Grund] for the fast growth, and the air, insofar as it is warm, is the cause [Ursache]; the fast growth, however, is grounded in the warm air. (GM, §29)

---

cause' and 'higher cause' in different ways in his work: In the natural sciences, Leibniz often speaks of 'efficient causes, while 'first cause' or 'first higher cause' is mostly correlated with his concept of God.

[3] Wolff writes: "Although Leibniz did not address the difference between cause and sufficient reason anywhere, the use he made of this principle nevertheless illuminated the path for me, such that I arrived at a clear concept of this principle, which corresponds with the definition of sufficient reason that I put forward earlier (§56)." (Ont., §71)

[4] In the original Latin this is: "Per *Rationem sufficientem* intelligimus id, unde intelligitur cur aliquid sit."

The conceptual distinction between 'cause' and 'reason' [Grund], as well as their relationship to each other, is not clear here—Wolff does not explain why the warm air is the cause and the warmth is the reason. Furthermore, it is not clear wherein the distinction between the principle of cause and the principle of sufficient reason lies—both the warm air and the warmth are material properties. Although Wolff does not give a definition of cause, we can, for our purposes, deduce Wolff's understanding of cause, and its distinction from sufficient reason, from his criticism of the scholastics. He writes, ten years later, in the *Philosophia Prima, Sive Ontologia*:

> The scholastics have long used the axiom *that nothing is without a cause*. But this sentence is not to be confused with the principle of sufficient reason, because reason and cause are different in a very significant way, as we will see in the following. This is why the scholastics did not admit an effect without a cause in physics, but did admit it [the effect] without a sufficient reason. They recognised, for example, that there must be a cause for magnetic attraction by appealing to the magnetic force in the magnet. However, because they did not think that it was necessary that there is something through which this magnetic attraction is intelligible, they admitted that the magnetic attraction could exist without a sufficient reason. (Ont, §71)

This example makes clear that the cause produces the effect [magnetic attraction], whereas the sufficient reason signifies that the content of the effect must be *understood* through this cause, i.e. the sufficient reason has to explain how the power of attraction is constituted such that it produces the attraction (for further reading cp. Schierbaum, 2022). Thus, Wolff explicitly distinguishes reason [ratio, Grund] from cause [causa, Ursache], whereby the former makes the effect intelligible in its existence. Interestingly here, the principle that every effect has a cause is a different principle from the principle that every effect has a reason, as we saw at the beginning of the quote above. Wolff accepts both principles, but it is only the principle of sufficient reason which is to him a fundamental axiom of human cognition (Ont, §56). Through the distinction between cause and sufficient reason it becomes evident that sufficient reasons give an intelligible reason for the effect, rather than simply expressing that the effect was caused by something, for this intelligibility is what distinguishes "cause" and "sufficient reason" conceptually. Yet although 'sufficient reason' is thus defined by Wolff in distinction from 'cause', he does not provide an explanation for how 'reason' and 'cause' relate to each other. While Wolff's view on what a reason is develops significantly in the ten years between the *German Metaphysics* and his *Philosophia Prima*, as we saw above, the definition of cause and in turn its distinction from and relationship to sufficient reason remains unresolved.

Du Châtelet read Leibniz and Wolff in close detail and concurs with Wolff that 'cause' and 'sufficient reason' are two different concepts that must be distinguished clearly. Differently to Wolff, however, Du Châtelet introduces a clear definition of cause and elucidates the relationship between 'cause' and 'sufficient reason'. A cause, in her definition, is the principle of the actuality of something.[5] Yet this cause is not only responsible for the actuality of its effect; the cause also contains something by

---

[5] "Beings" to Du Châtelet can be possible beings or existing beings (Inst1742, §35). Existing, actual, and real are terms which Du Châtelet uses interchangeably. There are also beings to Du Châtelet which do not *exist*. The ideal beings, space and time, for example, do not exist in and of themselves,

which an intelligent being can understand how and why the actuality exists in this being. She writes:

> Thus, a cause [cause] must contain [contienne] not only the principle of the actuality of the thing of which it is the cause but also its sufficient reason by which an intelligent being [Être intelligente] can understand why this thing exists: for anyone who knows how to use their reason must not be content to know that such a thing is possible, and that it exists, but they must also know the reason why it exists [...]; thus in everything that exists there must be something by which we can understand why that which exists can exist, and this is what we call *sufficient reason*. (Inst1742, §9)

The cause is here the material principle which actualizes a certain phenomenon as its effect. However, the cause does not only bring about a certain actual thing materially; the cause also contains 'a sufficient reason' which corresponds with the faculty of 'intelligence' in an intelligent being and which enables this being to understand how this actual thing came into existence. We can clarify this as follows: If a falling tree blocks the flow of water in the river, but no intelligent being perceives this tree, the tree still causes the flow of water to stop. If an intelligent being comes to the river and sees that there is very little water where there was a lot before, this intelligent being looks for a cause because the lack of water in the river can be understood in terms of its cause. The lack of water in the river means something different to this intelligent being depending on the cause, i.e. it may be more concerning, for example, if the cause is a drought than if the cause is the fallen tree.[6] Hence, the lack of water, to the intelligent being, is not only an observation but something that is *comprehended* through its cause. Du Châtelet's understanding of 'sufficient reason' is based on the fact that existence is intelligible in the relationship between cause and effect. Importantly, a 'sufficient reason' thus lies 'in the cause', i.e. in the principle of actuality, but corresponds with the 'intelligence' of the intelligent being. Hence, without an intelligent being no understanding of the actuality of a being and no understanding of a 'sufficient reason'; and without 'cause' no actuality of a being and no 'sufficient reason'. Du Châtelet thus makes clear that the material principle of actuality (cause) is from a theoretical or philosophical perspective not the same as the principle of the intelligibility of actuality (the sufficient reason). They do, however, stand in an inherent relationship to each other, insofar as a sufficient reason lies in the cause, i.e. a 'sufficient reason' is the principled potential for the cause to render the effect comprehensible to an intelligent being.[7]

---

but are nevertheless beings, insofar as their ideality makes a difference for what the world is, i.e. it can thereby be understood in a spatial and temporal order. The same is the case for numbers. Numbers do not exist, but they are beings and they make a difference for the comprehension of the world. They enable me, for example, to buy nine ice-lollies for nine children.

[6] We can view this more clearly using this example: If the death of a person is caused by a violent attack we say the death is a murder. If the death of a person is caused by an illness we say it was a 'natural' death. Hence, the comprehension of the cause of something affects the comprehension of what this something, i.e. the death, signifies.

[7] Du Châtelet writes: "Thus, a cause must contain not only the principle of the actuality of the thing of which it is the cause but also its sufficient reason by which an intelligent being can understand why this thing exists [...]." (Inst1742, §9).

In Du Châtelet's elucidation of the relationship between the concepts cause and sufficient reason we can see that the sufficient reason must be something which renders the actuality of a being *comprehensible*, i.e. the actuality of something must be explicable through its relationship to the particular cause given. I might for example say that in the example above not only the fallen tree is causally relevant to the lack of water further down in the valley, but the fact that water flows downwards is also causally relevant for the flow of water to have stopped. The fact that water flows downward is thus causally relevant for the effect, however the fact that water flows downwards is not on its own a complete sufficient reason for the actuality of the lack of water in the river because it does not on its own make it comprehensible why there is a lack of water: the fact is causally relevant but is only an aspect of the sufficient reason because this causal aspect, on its own, is not sufficient to make the lack of water comprehensible. The point of 'sufficient reason', then, is that it is that which makes the existence of a thing comprehensible through an understanding of its relation to all the material causes which brought this particular existence about. According to Du Châtelet, there must be an intelligible connection between all the causal factors (which together form *one* sufficient reason for an intelligent being) and the existence of a thing: and vice versa the assertion of causes which do not make the existence comprehensible as its effect is not a sufficient reason. Du Châtelet writes in this sense:

> But once it has been established that a cause is acceptable only insofar as it satisfies the principle of sufficient reason, that is, insofar as it contains something by which we can show how and why an effect can occur, then we can no longer accept these grand words that were used in place of ideas [by the Scholastics]. (Inst1742, §10)

We can thus summarize that in Du Châtelet the 'cause' is responsible for the material actuality of the existence of a being, while the 'sufficient reason' corresponds with intelligence. A sufficient reason lies in the cause because there is a principled potential in the cause to make the existence it brings about comprehensible to an intelligent being. For something to qualify as a 'sufficient reason' it must make the existence of a being *comprehensible* with a view to the cause (or to the causes in the plural). Nothing in this account contradicts either Leibniz's or Wolff's account of 'cause' and 'sufficient reason'; however, Du Châtelet provides us with a more detailed philosophical analysis of the concept of 'cause' and the concept of 'sufficient reason', as well as their relationship. Furthermore, the role of 'intelligence' in the comprehension of a 'sufficient reason' is thereby more precisely distinguished from the idea of a material relationship between cause and effect. Lastly, Du Châtelet's emphasis that in all that exists there must be something by which *we* can understand why it can exist makes clear that to her a 'sufficient reason' necessarily corresponds to *human* intelligence: the flow of water in our example cannot be stopped for no reason at all or, which would be the same for us, by a cause that is in principle unintelligible to us.

3.3 Du Châtelet's Theory of the Knowledge of Beings in their Essential, Attributive and Modal Determinations and the Epistemological Role of Sufficient Reason

In Inst1742, §9 Du Châtelet defines a sufficient reason as something "by which an intelligent being can understand why this thing becomes actual from having been possible, as it was before." We can derive from Du Châtelet's writings that to her God and the human being are intelligent beings (Inst1742, §1; §23). Hence, the 'sufficient reason' for any existing being corresponds with the intelligence of God and/or the human being and makes the existence or actuality of a thing or being comprehensible.

In Leibniz, a sufficient reason also corresponds with 'intelligence', i.e. with God and the human being as intelligent beings. Leibniz thought that any sufficient reason must be fully intelligible to God, while the intelligibility of sufficient reasons to the human being is a much more complex issue in his work (cp. Carus, 2025, Rutherford, 1992). While Du Châtelet shares Leibniz's idea that a sufficient reason corresponds with our or God's intelligence, her account is more theoretically explicit about what it is that is intelligible to the human being with a view to sufficient reasons (cp. also Amijee, 2024; Carus, 2025; Wells, 2021). Importantly, for Du Châtelet, the question of how, in what sense and to what extent existence is intelligible *to us* through the principle of sufficient reason is of crucial concern for what it is that qualifies as a 'sufficient reason'.

We already saw in Sect. 3.1 that a sufficient reason for Du Châtelet lies in the cause of the existence of something and giving the reason for the existence of something allows us to understand the relationship between the effect and the cause. In Chap. 3 of the *Institutions Physiques,* Du Châtelet develops, in a novel form, how attributive and modal determinations of a being are known on the basis of a sufficient reason. This theory stands in the context of her "doctrine of essences", in which Du Châtelet elucidates how beings are understood with a view to their essential, attributive and modal determinations.[8] According to Du Châtelet, the essential determinations make a being possible and remain the same even if a particular being with this essence seizes to exist. Hardness, for example, is an essential determination of a rock, and remains an essential determination of a "rock" even if a particular rock seizes to exist. Attributes and modes, on the other hand, pertain to the actuality of a being: the shape of a particular rock, for example, is constant in this particular rock while it exists but does not remain the same if that particular rock is gone. Similarly, the warmth of the rock is actual in this rock and is no more if it is not actual. We will first investigate the meaning of the attributive determinations of being in Du Châtelet before looking at the modal and essential determinations. This will allow us to take a step further towards our aim of showing that Du Châtelet develops the concept of

[8] Essential determinations are just 'essences', attributive determinations are just 'attributes', and modal determinations are just 'modes', but Du Châtelet's expression 'determinations' makes clear that beings are determined in different ways by their essential, attributive and modal 'determinations'.

sufficient reason further by determining it epistemologically, and in this context of ascertaining whether the postulation of God as a sufficient reason for the world is admissible.

3.3.1 Attributive Determinations of Beings and Their Comprehension

Attributive determinations of a being, according to Du Châtelet, are grounded in the essential determinations of a being and are recognized by the human being through their grounding-relation to the essence. Attributes thus have their 'sufficient reason' in the essence. Du Châtelet writes:

> [...] the properties or attributes have their sufficient reason in the essential determinations; for since once these essentials are posited the properties also are, one can understand by the nature of essential determinations why the attributes or properties are as they are, rather than otherwise. (Inst1740eB, §42)

To clarify this account of the intelligibility of attributes through their 'sufficient reason' in the essence, it makes sense to look at an example. My arm is understood through its ground in me as an essence and cannot be understood in its function as an arm without this grounding in me or my whole body – if I lose my arm the arm is not the same anymore because it is no longer an attribute of mine, does not have the same function etc. Du Châtelet's idea that attributes or properties have their'sufficient reason' in the essence of a being and are comprehensible through this grounding-relation is already present in Wolff. Wolff writes: "That which is grounded solely in the essence of a thing is called a property. Thus, the properties of a thing cannot be separated from a thing and are, just like the essence itself, invariable" (GM, §44, cp. also Ont, Part I, Sect. II, Chapter III, §157). Similarly, Leibniz claims that qualities of a subject are always explicable in terms of the nature of the subject. Leibniz writes:

> Whenever we find some quality in a subject, we ought to believe that if we understood the nature of both the subject and the quality we would conceive how the quality could arise from it. So within the order of nature (miracles aside) it is not at God's arbitrary discretion to attach this or that quality haphazardly to substances. He will never give them any that are not natural to them, that is, that cannot arise from their nature as explicable modifications. (A VI, 6, 66; for further reading see Rutherford, 1992)

Yet, as we can see, Leibniz's theory of the explicability of qualities in terms of their relation to the subject draws on a theological premise regarding the way God has created the world, while in Wolff and in Du Châtelet the theory of how attributes are to be understood rests solely on an analysis of the human comprehension of attributes in terms of their being grounded in the essence and their being comprehensible through this grounding relation. In contrast to Leibniz's theological postulates, Wolff calls the first chapter of his *German Metaphysics*, in which he is concerned with the grounding of attributes in essences through 'sufficient reason' "Of the Grounds of our Knowledge" and does not reference God at all in the context of attributes (GM,

Chap. 1, cp. also Ont, Part I, Sect. II, Chapter III). For Du Châtelet too the conceptual understanding of what an attribute is is grounded in epistemological considerations concerning their relationship to essences in the human knowledge of beings, rather than in any theological premises. In this context, she writes: "Everything that is deduced from the essence is constantly apparent in a being and is what we call *attribute* or *property*" (Inst1742, §39). And "[...] one can understand through the nature of essential determinations why the attributes or properties are as they are [...]" (Inst1740eB, §42). In other words, we can note a shift in Wolff and Du Châtelet away from theological premises of an intelligibility that is attributed to God (which is only secondarily attributed to the human being) towards taking the intelligibility of human reason as the main premise for this philosophical argument (cp. also Wells, 2021, 636f.). In Leibniz, by contrast, there is a lineal succession of sufficient reasons that is only comprehensible with a view to his conception of God. For example, Leibniz writes in the *Discourse on Metaphysics*, §7:

> It is indeed true that when several predicates are attributed to the same subject and this subject is not attributed to any other, it is called an individual substance. But this is not sufficient, and such an explanation is only nominal. It must therefore be considered what it is to be attributed truly to a certain substance. [...]Thus, the quality of King that belongs to Alexander the Great, taken in abstraction from the subject, is not sufficiently determinate for one individual, and does not include the other qualities of the same subject, nor does it include everything that the notion of this prince contains, while God, seeing the individual notion or haecceity of Alexander, sees in it at the same time the ground and the reason of all the predicates that can be said truly of him [...] even knowing a priori from it (and not by experience) whether he died a natural death or by poison, which we can only know through history. (For further reading see Melamed/Lin, 2023 and Rutherford, 1995)

We can see clearly here that for Leibniz attribution of substances is only sufficiently explained through recourse to God—without this recourse a philosophical explanation of attribution is nominal and insufficient. This is decidedly not the case for Wolff and Du Châtelet, for both of whom the philosophical theory of attributes and attribution is fully based in an epistemology, i.e. principles that are fully intelligible to the human being. This will, as we will see, have consequences for what can be posited as a sufficient reason.

3.3.2 *Modal Determinations of Beings and Their Comprehension Through Sufficient Reason*

Turning now to the intelligibility of modes through 'sufficient reason', Du Châtelet writes: "The simple possibility of modes has its sufficient reason in the essence; but their actuality depends either on other antecedent modes, or on exterior beings, or on both at the same time" (Inst1742, §44). Let us elucidate this universal intelligibility of modal determinations to the human being more closely: A human being's essence is such that a fever is a possible mode of his being, while a circle's essence is such that a fever is not a possible mode of this being. Hence, there is a relation between the essence of a being and the possibility of the modes to which it is susceptible. However, the actuality of a mode is comprehended through a different relation: I can, for example, understand the fever a person has (a mode) by understanding the cause in either an exterior being (a virus) or in a previous mode (hypothermia). If I know what it is that brought about the fever, I understand the fever in its actuality or existence – I know why it is there, i.e. I know its sufficient reason. The fever is understood through this sufficient reason such that an intelligent being can know how concerning it is, how it has to be treated etc. In other words, by virtue of the knowledge of the sufficient reason for the fever given in the cause or causes, the fever becomes an object of medical knowledge.

Du Châtelet's account of this intelligibility of modes through their 'sufficient reason' in exterior beings or antecedent modes is a theory of causality. In the sequence of things there is always a sufficient reason why the modes of a being are as they are, and this sufficient reason is intelligible for human beings. In this sense, a human being might not know the sufficient reason for the fever but there is a cause which brings the fever about, and this cause is a sufficient reason which has the potential to make the existence of the fever comprehensible.[9]

3.3.3 Essential Determinations of Beings, Their Comprehension and Their Unintelligibility with Regard to Sufficient Reasons

In this context, the question arises whether we can know the essential determinations of a being and in what sense. Du Châtelet emphasizes that the essential determinations of a being have no further first and intrinsic reason [raison]. Insofar as the essential determinations are the first thing that we recognize in an entity, these cannot in turn be traced back to another reason (Inst1742, §41). In this sense, Wells (2023, 47) concludes that Du Châtelet is pessimistic regarding human knowledge of the essence of existing things, because she thinks that the principle of sufficient reason does not apply to essential determinations. He writes:

> Her PSR [in Du Châtelet] ranges over contingent truths, however, and not over necessary truths about the essences of finite things. Therefore, her PSR is silent about our knowledge of necessary properties of the essences of finite things. Given that the PSR is closely linked to her conception of science, it follows that her account of science does not regard it as primarily aiming at knowledge of essences. In contrast, we saw that Wolff takes science to truly describe

[9] Wells (2021, 633) writes, with a view to the distinction between actually knowing a particular sufficient reason and the principle possibility of knowing this reason: "Note that Du Châtelet's PSR is modalized: it does not guarantee that we will in fact understand the sufficient reason for any given contingent truth, but only that an intelligent being "could understand [*puisse comprendre*]" the reasons in question (IP I.9/ZB 131; translation modified)." Amijee (2024) terms Du Châtelet's theory of causality based on the fundamental knowability of natural reasons "epistemic causal idealism". She writes of Du Châtelet's theory in this sense: "if a cause exists in the natural world, then it is possible for us to know it - natural causes just are possible objects of knowledge.".

> essences and does so in part because of specific features of his PSR. Appreciating this feature of Du Châtelet's PSR is crucial for reconciling her version of rationalism with her pessimism about our knowledge of essences.

Although I agree with most other points of Wells's paper, I am doubtful about the conclusion that Du Châtelet is 'pessimistic' regarding our knowledge of essences. Du Châtelet writes:

> [...] the possibility of things is not dependent on his [God's] will: because the essences of beings consist in their first determinations, the combination of which makes them possible (§ 38); this essence is not dependent upon the will of God, instead it has its fundament in the principle of contradiction.[10] (Inst1742, §47)

Du Châtelet continues with this exposition by explicating how the essential determinations of a triangle and of a rock are deduced on the basis of the principle of contradiction. It is clear from her exposition that we can know what the essences of beings are. According to Du Chatelet, knowing a being in its essence signifies that we know all the constant determinations in this being, which are not dependent on any other determinations, like the attributes are. If we know these "first" determinations of a being and know that they do not contradict each other, we know its essence. According to Du Châtelet, our knowledge of essences is grounded in the PC: we can decipher whether something truly is a being by investigating whether its first determinations, which are not dependent on anything else, are contradictory or not (Inst1742, §38). Hence, although Wells is right that the PSR does not reign over essences in Du Châtelet, she thinks that we can know essences, i.e. we can know *what* they are, but we cannot know *why* they are what they are. In fact, it is constitutive for Du Châtelet's definition of what an essence is that a "sufficient reason" cannot be provided: if a constant determination of a being is grounded in yet another determination of this being, i.e. is comprehensible through a different determination, then it is an attribute, not an essence. Essences and attributes are both constant and distinguished solely by the fact that essences are only comprehensible in themselves, i.e. are not grounded in another determination, while attributes are grounded in essences and are comprehensible through their relation to the essence.

The comparison between Du Châtelet's and Wolff's theory of essences is interesting. In the *German Metaphysics* of 1720 Wolff claims that any essence has a sufficient reason (for being) in itself by means of the PC (GM, §29–31). 10 years later, in the *Philosophia prima, sive ontologia* Wolff changed his position and writes "because essences are infinite, no intrinsic reason for essences can be given" (Ont, §156). Yet essences can be known, by virtue of the PC, if we know their "intrinsic possibility". Thus far, Wolff's revised theory of essences sounds similar to Du Châtelet's and she quotes it as a reference in the preface of the 1740 edition (cp. Inst1740, XII; curiously the footnote with the reference to Wolff's Latin *Philosophia prima, sive ontologia (Ontologia)* is missing in the 1742 edition). However, Wolff's theory of essences in the *Philosophia prima* still has traces of his earlier view that essences have intrinsic reasons. Wolff, namely, still holds that we can only know the "intrinsic possibility"

[10] Cp. also Inst1742, §50 and §27.

of essences, and in turn the essences themselves, if we can show a priori how this essence "was produced" (cp. Ont, §154, §155). Wolff's position is thus still that knowledge of essences is gained through knowledge of the way they "come about", only now he does not call this a "reason", as he did in the *German Metaphysics,* and instead calls it a "*definitionem geneticam*". We know the essence of a watch, according to Wolff, when we know the structure that enables the watch to be what it is (Ont, §155). Du Châtelet's point, when she says that essences do not have a reason, is, I think, slightly different to Wolff's: we can comprehend a being in its essence, but it is unintelligible to us why the essence of a being is the way it is – and this is more important than one would at first think. We can, for example, know why a person exists with a view to the order of things in its attributes and modes, i.e. the parents etc. and we can understand what a person is in essence by knowing all the determinations in this person (this being), which are constant and do not depend on any other determinations, but we cannot understand why this person in their essence is what it is.[11] We can know that the human being is rational, but not why it is such that the human being is rational: we can imagine that a higher power had some intention in creating beings in their essences in the way they are, but we do not know this because the cause of the essences of beings, how the essence of beings came about, is not intelligible to us. When we understand the essence of a watch by means of its internal structure we do not understand how the essences of the watch were produced - instead, we understand how the being, i.e. the watch, is possible. Du Châtelet's point when saying that essences do not have a reason is, then, in contrast with Wolff, that essences are not intelligible with a view to how they come about, come to be, or are "produced" (cp. also Du Châtelet's comment on Voltaire's *Traité de metaphisique* in the St. Petersburg manuscripts (PM, 11)). In Du Châtelet, we know the essence of a being when we know the determinations that are constitutive of that "being" [être] being what it is (cp. Inst1742, §38), not by knowing the intrinsic possibility of the essences themselves. In contrast to Wolff, knowledge of essences in Du Châtelet is not dependent on knowing how they can be, but on knowing *what* the essential determinations of a particular being are.

We have seen in this section that the actuality of attributes and modes are fully intelligible to the human being on the basis of 'sufficient reason' and that Du Châtelet expounds in detail how this intelligibility is to be understood in philosophy. Her theory of the intelligibility of attributes and modes through a sufficient reason does not, in its argument, depend on any theological premises of Godly creation. Du Châtelet establishes an epistemological theory of how beings are comprehended by the human being in terms of a sufficient reason and with regard to the limits of the human comprehension of beings in terms of a sufficient reason: Beings in their modal and attributive determinations are intelligible through the causes of the mode and through the grounding of the attribute in the essence, but essences of beings are not

[11] Du Châtelet writes: "There is no further primary and intrinsic reason why the essential determinations of a being are found in this being; because these determinations are the first which one can conceive in a being we cannot conceive of something anterior on which these determinations themselves depend." (Inst1742, 41).

intelligible through a further ground or cause, for they are the first thing through which we conceive of a being. Hence, beings are understood in what their essence is but the problem is "overdetermined" and the question posed wrongly if we ask why beings in their essence are what they are or how they came about, for this "cause" of essences is not intelligible to us.[12]

3.3.4 The Intelligibility and Unintelligibility of the Existence of Beings

We saw above that in her 'doctrine of essences' Du Châtelet establishes that there is no first or intrinsic reason for why the essential determinations of a being can be found in a being (Inst1742, §41). Like Wolff, Du Châtelet holds that the essential determinations are the first determinations we conceive of in a being and concludes from this that we cannot conceive of anything antecedent to the essence. However, differently to Wolff in the *German Metaphysics* she does not conclude that the essential determinations have a sufficient reason [Grund] in themselves (GM, §29–31). Rather, she thinks this cannot be the case because sufficient reason signifies that we understand the thing in question through a relation to a cause in something else. Du Châtelet's view that essences do not have a reason has its source in Wolff's *Philosophia prima, sive ontologia.* However, differently to Wolff in the *Philosophia prima,* Du Châtelet thinks that essences are not intelligible with a view to how they "are produced". Rather, essences are given in the beings that we cognize, and we can know essences by asking ourselves what the determinations of the being are that are permanent, independent of other determinations, and make the being what it is.

Du Châtelet establishes in Inst1742, §48 that we can consider the existence of a being in two distinct ways: we can ask ourselves why any given being is actual or existent, considered in itself, or we can ask why it exists with respect to its position in a sequence relating to other beings. While the latter is fully intelligible, as we saw above, the existence in the former sense is unintelligible; the human being, according to Du Châtelet, cannot know why the essence of a being is what it is or why creation, i.e. all beings taken together, is what it is.

Let us spell out the difference with a view to Du Châtelet. We saw above that in terms of the actuality of beings their attributes are intelligible with a view to their essence, and the actuality of a being's modes is intelligible in a sequence of beings

---

[12] In Du Châtelet's metaphysics the principles and concepts, which are employed naturally in the human comprehension of beings, are made theoretically explicit. Du Châtelet writes: "This principle, on which all contingent truths depend, and which is neither less primitive nor less universal than that of contradiction, is the principle of sufficient reason: everyone follows it naturally; [...]." Du Châtelet thinks that metaphysics is the "foundation of all knowledge" (Inst1742, §1) and the "summit of the edifice" (Inst1742, XII) in precisely this sense: human knowledge rests on metaphysical principles and concepts and in metaphysics these become theoretically explicit; the explication of which is the zenith of human knowledge.

with a view to a previous mode or an exterior being. It is this intelligibility that allows beings to be understood in philosophy, according to Du Châtelet. She writes:

> Thus when it is a question of admitting any properties in a being, one must see if this property is grounded in its essence, that is to say, in the first determinations which make it possible; for when a being is considered alone, one must show its intrinsic possibility by the principle of contradiction, and its external possibility, or its actuality, by the principle of sufficient reason, and deduce the attributes of this being, and the modes to which it is susceptible through it. And when one considers this being as placed in the sequence of things, [la suite des choses] and linked with the other beings that surround it, one must show how a being depends upon other beings, and which causes gave actuality to the modes that were simply possible before when one considered the being in isolation and outside the sequence of things. It is in this way that God executed his will, and that one must seek to give reason to things in philosophy. (Inst1742, §48)

In other words, the actuality of a being is intelligible with a view to the attributes of a being and the modes of a being since both have a cause or reason in something else and that relation is intelligible to the human being. Furthermore, the essences of beings are intelligible in what they are, but they do not have a cause or reason in yet another determination: they are primordial. The idea that God has created the world in this intelligible way plays no constitutive role in Du Châtelet's argument: beings simply are intelligible to the human being in their attributes and modes through a relation to a cause or reason, i.e. on the basis of giving a sufficient reason for their existence, and beings are intelligible in their essence through any being's essential determinations being non-contradictory.

However, in the second chapter of the *Institutions Physiques* a precarious idea is brought to bear on the sufficient reason for beings. Du Châtelet explains that the actuality or existence of beings in the sense of creation as a whole is a different consideration to the actuality of a being's attributes or modes - and depends on God's will. Du Châtelet begins with the premise that, from a human perspective, the entire chain of existing beings could be different to the way it actually is. Hence, the intelligibility of existence in terms of a sufficient reason, as described above in terms of attributes and modes, does not allow the human being to know where the determination of the entire chain comes from, i.e. why the entire infinite chain of existence is the way it is. She chooses the analogy of a novel to make clear that the things that exist could also have been different to the way they are:

> In a word, he [God] could take an infinity of courses of action different from the one he took, which would have given rise to an infinity of combinations that were then possible and that would have produced events all different from those that occurred. This applies to the events contained in novels. They could happen if another succession of things took place; these are the stories of a possible world that lacks actuality, for each succession of things constitutes a world differing from all others by the events specific to it. Thus one can conceive of a succession of causes leading to the events in Zaïde, or those in the Queen of Navarre, for these events are possible, and they only lack actuality. Similarly, one can conceive of possible universes, with other stars and other planets; and, as the different relationships of these universes can be combined in an infinity of ways, there are an infinity of possible worlds, only one of which actually exists. (Inst1740eB, 23)

We can see here that Du Châtelet, on the basis of the concept of God, imagines that a different course of the actualities of things could have ensued, i.e. that the world could have been different to the way it is. The erroneous use of the concept of sufficient reason occurs in connection with this imagination of other possible worlds, as we will see, when Du Châtelet claims that there is a sufficient reason for the existence of this world in the way that it is in God's will. According to Du Châtelet, the actuality of things lies in God's will. She writes: "The choice of the creator is the sufficient reason for the existence of this world" (Inst1742, §25). Now, according to Du Châtelet, this choice of existence by the creator, i.e. why all beings exist as they do and why every singular being is connected to other beings in the way that it is, is not accessible to human understanding because the human being cannot conceive of all the possibilities. She writes:

> The way God sees and imagines all sorts of things is incomprehensible to us. Therefore, we cannot form a clear idea of the divine mind; like creation, it is one of the things that we can neither understand nor deny. (Inst1742, §24)

We can see here that creation, i.e. the entire sequence of contingent things in their successive reality, has a postulated sufficient reason; however, we can 'neither understand nor deny' this reason. Du Châtelet is as clear as can be in saying that creation as well as God's reason for creation is beyond human cognition.[13] We simply do not know why the creator chose to write the story of the world the way it is, and not differently. It would follow from this that we cannot postulate a Godly or moral reason for why beings are created in the way they are, for such a reason is simply not intelligible to us.

Inconsistently with a view to her own doctrine of essences and her realization that creation cannot be understood in terms of sufficient reasons, in the way attributes and modes can thus be understood, Du Châtelet still abides by the theological doctrine of her predecessors and postulates that there must be an (unintelligible) sufficient reason for the entire infinite chain of existence and furthermore that this sufficient reason brings about the determination of existence as a creation of the best of all possible worlds. She claims that the fact that all existing beings come into existence as they are and not otherwise has its sufficient reason in an intelligent being that conceives of all possible worlds in his understanding and chooses the one that is best.[14] However, the theological positing of such a sufficient reason, which is unintelligible to the human being, is in clear tension with her own elucidation of what a 'sufficient reason' is. From her own account of the necessary intelligibility of sufficient reason to the human being it would logically follow that a sufficient reason posited for the existence of the world (God) and for the existence of everything in the way it is (the best of all

[13] The roots of this idea lie in Leibniz who also thinks that the sufficient reason for creation is 'above' human reason, i.e. does not contradict human reason but is also not intelligible to it (GP VI 64; for further reading Antognazza, 2013). However, in Du Châtelet this amounts to a stronger tension because her elaboration of the principle of sufficient reason implies that sufficient reasons are per se intelligible.

[14] For Du Châtelet this amounts to a proof of the existence of God (Inst1742, §19, for further detail see Lascano, 2011).

possible worlds) is a contradiction in terms. Of course, we can imagine and believe that the world is grounded in an intelligent creator, however, this is a matter of faith, not science. How can we posit that there is a sufficient reason for creation and for beings being the way they are in their essence if this sufficient reason is[15] entirely unintelligible to the human being?

Despite the problematic adoption of the theologically motivated postulation of the existence of God and the doctrine of the best of all possible worlds, we must recognize that Du Châtelet is the first Early Modern author to clearly assert that the question 'why a being is', taken in itself, is a different question to why the attributes of a being are what they are or why the modes, taken in their position in the sequence of things, are what they are. Furthermore, while she still postulates that the essences of beings are "not outside of God" and rooted in "God's understanding", she makes clear that we cannot postulate a "sufficient reason" for essences because the "cause" or "reason" why beings in their essence are the way they are is not intelligible. Beings in their essence are fully comprehensible but asking after the "cause" or "reason" for why the essences of beings are the way they are is unphilosophical, unintelligible and overdetermining.

3.4 Conclusion

We saw that Du Châtelet develops the understanding of what it is that constitutes a 'sufficient reason' significantly further. Firstly, the relationship and distinction between 'cause' and 'reason' was elaborated *conceptually* by Du Châtelet. While Leibniz does not strictly distinguish between the two concepts, Wolff does not explain what role the cause plays for the determination of what a sufficient reason is. The relationship between cause and reason is only made explicit by Du Châtelet. She not only gives a definition of sufficient reason but also of cause and explains how the two concepts relate to each other: Sufficient reason, for her, signifies that the cause, which contains the principle of the actuality of a being, can be known and that the actuality of the being in turn can be understood, i.e. is an object of knowledge.

Secondly, the intelligibility of attributive and modal determinations of a being in terms of a sufficient reason is further elaborated on in Du Châtelet. In the philosophical elaboration of what modes and attributes are and in establishing how they are to be principally comprehended we can note a shift away from Leibniz's theological premises regarding how God has brought about the constitutional elements of

[15] Clearly, the idea of a cause for causation can be posited in terms of a question relating to a series that asks after an initial cause that is the reason for the causation – this idea is not itself contradictory. However, the being in its essence or mode or attribute, from which we derive the idea of an absolute initial cause in the series of causes, cannot possibly be explained in its content by the idea of an absolute initial cause because the question of an initial cause is not related to the real being in its "essence" or "attribute" or "mode", but rather to an idea relating to the series of causes as an idea that has nothing to do with sufficient reason.

the world. Furthermore, in distinction from Wolff, Du Châtelet makes clear that the essences of beings are not intelligible to us with a view to how they came about.

Du Châtelet's contribution to the concept of sufficient reason is that sufficient reason signifies that existing beings are intelligible with a view to a cause or reason in something else: modes are intelligible through their relation to a previous mode or to an exterior being that caused the mode, and attributes are intelligible through their relation to the essence. Existence is universally intelligible in this way: no singular being, no mode and no attribute, exists without being caused by something else or having a reason in something else and thus we can always give a sufficient reason for the existence of beings. Beings in their existence become an object of knowledge through giving a sufficient reason for their existence. Why beings in their essence are what they are is not intelligible, hence we cannot give a sufficient reason for this. If Du Châtelet had been consistent in this convincing epistemological determination of what sufficient reason is, she would have used the concept sufficient reason only to signify that any being's existence is intelligible solely with a view to its relation to a reason or cause that explains this being's existence. However, her use of the concept, sufficient reason' in the context of God's creation of the world, which is unintelligible, contradicts her own determination of what a sufficient reason is. Nevertheless, in her work and in view of this contradiction, the question as to whether a sufficient reason, which is unintelligible to the human being, can be legitimately posited becomes more pressing than in Leibniz or Wolff. Is an unintelligible sufficient reason not a chimera?

References

Amijee, F. (2024). Du Châtelet's Causal Idealism. *British Journal for the History of Philosophy,* 1–22.

Antognazza, M. R. (2013). Faith and reason. In M. R. Antognazza (Ed.), *The Oxford handbook of Leibniz* (pp. 717–734). Oxford University Press.

Bobro, M. (2021). Leibniz on causation. In E. N. Zalta (Ed.), *The Stanford encyclopedia ofphilosophy.* https://plato.stanford.edu/archives/fall2021/entries/leibniz-causation/. Last accessed November 2023.

Carus, C. (2025). *Das Prinzip vom zureichenden Grunde bei Émilie Du Châtelet. Eine philosophiegschichtliche Einordung von Du Châtelets Beitrag zum Prinzip vom zureichenden Grunde in einem Vergleich zu G.W. Leibniz und Christian Wolff.* Springer.

Lascano, M. (2011). Emilie du Châtelet on the existence and nature of god: An examination of her arguments in light of her sources. *British Journal for the History of Philosophy*, 741–758.

Melamed, Y. (2014/2023). The principle of sufficient reason. *Stanford Encyclopedia of Philosophy.*

Pikkert, O. (2018). *Leibniz and the principle of sufficient reason.* Ph.D. thesis, Toronto. https://tspace.library.utoronto.ca/bitstream/1807/102922/3/Pikkert_Owen_201811_PhD_thesis.pdf. Last accessed December 02, 2023.

Rutherford, D. (1992). Leibniz's principle of intelligibility. *History of Philosophy Quarterly, 9*(1), 35–49.

Rutherford, D. (1995). *Leibniz and the rational order of nature.* Cambridge University Press.

Schierbaum, S. (2022). Christian Wolff über motivierende Gründe und handlungsrelevante Irrtümer. *Archiv Für Geschichte der Philosophie, 104*(1), 131–163.

Wells, A. (2021). Du Châtelet on sufficient reason and empirical explanation. *The Southern Journal of Philosophy, 59*(4), 629–655.

Wells, A. (2023). Science and the principle of sufficient reason: Du Châtelet contra Wolff. *Hopos: The Journal of the International Society for the History of Philosophy of Science*, 24–53.

Clara Carus is Alexander-von-Humboldt Postdoctoral Fellow at Harvard University. Her research focus is on early modern metaphysics, especially the works of Leibniz, Wolff and Du Châtelet.

Chapter 4
Du Châtelet on External-To and the Idea of Extension

Qiu Lin

4.1 Introduction

Du Châtelet's *magnum opus, Foundations of Physics* (1740 and 1742), contains a chapter entitled 'Of Space,' in which she offers her response to the heated debate between Newtonians and Leibnizians regarding the nature of space.[1] Here, she states explicitly that her goal is not just to refute Newtonians on metaphysical grounds. More importantly for her, she wants to demonstrate why it is tempting to make the kind of mistakes that Newtonians make. One such mistake Newtonians make, according to Du Châtelet, is to hold that space is 'absolute' in the sense that it is a substance-like entity existing independently of the bodies in it. She demonstrates the root of this error by examining how we form the ideas of space and spatial properties (e.g., continuity, infinity, voidness, and penetrability). Instead of staking out her position in the ongoing debate about the metaphysics of space, that is, Du Châtelet's chapter is far more concerned with the epistemology of space.

Before turning to address conceptions of space and spatial properties, however, Du Châtelet examines our idea of extension, asking first how we arrive at this idea. She dedicates two lengthy paragraphs to answering this question. According to Du Châtelet, the idea of extension necessarily arises when we 'represent to ourselves several different things as being one' via the imagination (§77). Moreover, in order to represent things as 'several' (i.e., as numerically distinct), 'in our minds we place one external to the other' (§77). But what does it mean to place one thing externally to another in our minds? This question has been the focus of much recent scholarly debate on Du Châtelet's work, which falls roughly into two categories. On the one

[1] All passages cited in this chapter remain unchanged in the 1742 edition, which, according to Du Châtelet, was 'corrected and augmented considerably' over the 1740 edition.

Q. Lin (✉)
Simon Fraser University, Burnaby, BC V5A 0C8, Canada
e-mail: qiul@sfu.ca

C. Carus and J. K. McDonough (eds.), *Émilie Du Châtelet in Relation to Leibniz and Wolff*, Women in the History of Philosophy and Sciences 23,
https://doi.org/10.1007/978-3-031-80541-7_4

hand, Katherine Brading's position is that 'external' is to be understood as a spatial relation (Brading 2019, 61). According to this reading, by placing one thing external to the other, our minds represent an object as spatially outside of another. On the other hand are those who interpret Du Châtelet's terminology as referring to an ontological relation, whereby our minds represent one as ontologically independent of the other. Following in the footsteps of the scholars who first noted this distinction, I shall refer to these perspectives on Du Châtelet's use of 'external' respectively as the 'spatial reading' and the 'ontological reading' (Jacobs, 2020, 68).[2]

This chapter explores this controversy in depth and offers a defense of the spatial reading along the way. Before doing so, however, it is worth explaining why it is so important to scrutinize the term 'external' in such detail. In brief, the term 'external' is key to understanding Du Châtelet's origin story of how we come by the idea of extension in §77; and, in turn, understanding §77 is essential if we want to make sense of both Du Châtelet's account of how we come by the idea of space and her theory of monads. The importance of the debate surrounding 'external', therefore, extends far beyond its own contents. After establishing the importance of this debate, we shall carefully clarify its scope and introduce the two competing readings of 'external' in the debate, the ontological reading and the spatial reading. Finally, three problems will be raised against the ontological reading: the Problem of Historical Precedent, the Problem of Mismatch, and the Problem of Extended Souls.

4.2 The Philosophical Importance of §77

'Of Space' touches upon a number of heated disputes regarding space that were circulating at the time. Du Châtelet's contemporaries sought to answer several questions, including: is space empty, or is it filled with matter? Is space 'absolute,' an all-encompassing container wherein bodies exist, or is it 'relational,' as the mere order of things as they coexist? Relatedly, are places 'absolute,' as parts of an encompassing void, or are they 'relational,' as particular manners by which objects coexist with other objects? With respect to the second of these questions, Du Châtelet expresses her clear approval of Leibniz's relationism as well as his two arguments against absolute space (one appealing to the Principle of Sufficient Reason and the other citing God's attributes), without adding anything new to either of them. She then changes tack to do something more interesting still: instead of focusing on the nature of space, she turns to investigate our ideas of space and spatial properties. This maneuver occurs at §77, which opens with the following remark:

[2] This debate was initiated by Jacobs (2020), who distinguishes between his proposed ontological reading and the spatial reading advocated by Katherine Brading and Marius Stan. See Jacobs (2020, 68): 'Indeed, recent commentators such as Stan and Brading subscribe to the strong reading without considering the weaker alternative'.

> {**How we form the idea of Space and its properties.**} It will not be without use to examine here how we came to form our ideas of extension, Space, and continuity; this examination will reveal to you the source of the mistakes that have been made about the nature of Space, and prevent you from making them in the future. (§77)

While the present chapter focuses on the external-to relation in Du Châtelet's work, this discussion has broader implications, as it will illuminate her later treatments of the ideas of relational space and absolute space. The reason for this is that understanding the external-to relation is crucial for understanding Du Châtelet's account of how we first acquire the idea of extension, which precedes her discussion of how we obtain the idea of space as the order of coexistence.[3] The organization of the text suggests that, for Du Châtelet, in order to understand how we arrive at the idea of space, we need to first explain how the mind acquires the idea of extension. Indeed, two paragraphs later, Du Châtelet argues explicitly that the idea of extension is a significant component of our idea of space: 'Thus, as long as we consider the possibility that several different things can exist together in this abstract Being we call *Extension*, we form the notion of Space' (§79).

In addition, she maintains that once we have grasped her argument about the relationship between these two ideas, we will recognize why Leibnizians 'are right to define Space ... [as] the order of coexisting things' (§79). Within her account of how we arrive at the idea of space, that is, lies her defense of the Leibnizian idea of space as the order of things as they coexist. And as we can see from carefully studying later paragraphs, she repeatedly returns to features of our idea of extension when she attempts to demonstrate that we tend to think of space as having (absolutist) properties:

> Thus, we are led to represent Space as a substance independent of the Beings that are placed in it (§80).
>
> Likewise, Space must appear to us to be void and penetrable. (§82) Space must also appear immutable to us. (§83) It must also appear to us infinite. (§84)

Her account of the provenance of the idea of extension, therefore, contains details that are crucial when we want to examine, on the one hand, whether her Leibnizian position on space holds and, on the other, whether she ultimately succeeds in demonstrating that Newtonians are wrong about the nature of space.

Moreover, the significance of §77 extends beyond the bounds of her chapter, 'Of Space'. §77 is, for example, cited in §133, which belongs in a different chapter altogether, 'Of the Elements of Matter'. Here, Du Châtelet seeks to establish a theory of matter underwritten by monads (also referred to as 'Simple Beings' and 'Elements'). She writes:

> For the Elements exist all necessarily external to one another (since one can never be the other), and all of them being, as we have just seen, united and linked together, the result

[3] At this point in the text, Du Châtelet already rejects absolute space as 'an illusion of the imagination' and endorses Leibniz's conception of space as the order of coexistence, claiming it as the prevailing position in the Leibniz-Clarke debate (§74). For more discussion about Du Châtelet's reception of this debate, see Hutton (2012, 88–89), Hagengruber (2011), and Reichenberger (2021).

> is an assemblage of several diverse Beings that exist external to one another and by their connection make a whole. *But I have shown that we cannot represent extension other than as an assemblage of several diverse, coexisting things, which exist external to one another* (§77): therefore, conclude the Leibnizians, an aggregate of simple Beings must be extended. (§133, my emphasis)

As this passage suggests, one purpose of §133 is to explain how it could be the case that, in experience, we only ever encounter extended objects. This observation would seem to fly in the face of a monadological theory of matter, whereby those beings that really exist are 'simple' in the sense that they do not have extension. Here, we find an essential link between Du Châtelet's metaphysics and her epistemology. The problem, as Du Châtelet states it, is a metaphysical one: it concerns how the non-extended constitutes the extended, a serious challenge facing every metaphysician who maintains that the most fundamental constituents of matter are, unlike matter itself, not extended. But Du Châtelet appeals to §77 before concluding the argument, where §77 is concerned with epistemology, examining the origin of our idea of extension. Du Châtelet's answer to a metaphysical problem therefore depends on her answer to an epistemological problem; more specifically, her answer depends on her account of how our minds form the idea of extension. Seeing this connection of her epistemology and metaphysics, she characterizes her argument as one that seeks to recover 'the mechanical union of the Bodies *that we see*' (§133, my emphasis). Thus, understanding §77 provides a lens through which we can see how the different parts of Du Châtelet's system hang together. Now, let us proceed to take a closer look at §77.

4.3 §77 and the Debate Surrounding 'External'

In §77, Du Châtelet invites us to consider three examples of how we ordinarily represent things: representing 'an imagined structure that we have never seen before', representing 'two persons', and representing 'the same person twice'. She writes:

> If in our imagination we represent to ourselves a structure that we have never seen before, we represent it as external to ourselves, even though we know well that the idea we have of it exists within us, and that there is perhaps nothing of this structure existing external to our idea. But we still represent it as being external to ourselves because we know that it is different from us; likewise, if we represent two men ideally, or even the same man twice, we place them external to one another, because we cannot force our mind to imagine that they are *one*, and *two*, at the same time. (§77)

The observation she makes from these examples is the following:

> We feel that, once we consider two things to be different, and when we distinguish one from the other, in our minds we place one external to the other; thus, everything that we consider to be different from us we see as external to us; there are many examples of this. (§77)

In these quotations, Du Châtelet speaks of how our minds 'distinguish' things, i.e. we do so by placing 'one external to the other'. In what sense should we understand these two key notions?

The examples of representing 'ideally two persons, or even the same person twice' should suffice to show that by 'distinguish', Du Châtelet means the making of a numerical distinction. In these examples, what readers are invited to ponder are the ways in which they ordinarily represent two distinct persons (Sarah and Michael, say) and two instances of the same person (two instances of Sarah, say) via the imagination, respectively. In the first case, Sarah and Michael are both qualitatively and numerically distinct. In contrast, in the case of representing two instances of Sarah, the two instances are qualitatively identical. Therefore, when Du Châtelet refers to representing 'the same person twice,' she means representing two qualitatively identical instances of a person. The distinction to be made is not qualitative but numerical.

How should we understand what Du Châtelet means by 'external'? Given her description of the examples, the most natural reading of 'external' would be spatial: in distinguishing two objects as numerically distinct from each other, Du Châtelet thinks, we place one as spatially external to the other in the sense that one occupies a distinct place from the other. However, a disagreement arose in a recent discussion about this term 'external'. As an alternative, Jacobs (2020) argues that 'external' is here denoting a *metaphysical* relation, thereby instigating a dichotomous framing of the following debate surrounding 'external' (a 'strong reading' vs. a 'weak reading'):

> On the strong reading, 'external' is a spatial notion. Consequently, we necessarily represent distinct simples as located in space at some distance from each other. On the weak reading, 'external' is not a spatial but a metaphysical relation such as ontological independence (i.e. if A is external to B, then A does not depend on B for its existence). [...] At first sight, the weak reading seems implausible: what can 'external' mean, besides spatial removal? Indeed, recent commentators such as Stan and Brading subscribe to the strong reading without considering the weaker alternative.[4] (Jacobs, 2020, 68)

In what follows, I offer a critical examination of the ways in which Jacobs set up the debate over interpreting 'external' in §77 of Du Châtelet's chapter on space. I intend to show two things. First, Jacobs's attribution of what he calls 'the spatial reading' to Stan is not supported by good textual evidence. This alleged contender in the debate did not weigh in on the issue being debated. Second, I point out the subtlety inherent in Du Châtelet's uses of 'external' for different relata, and delineate a suitable scope for the debate accordingly. A careful taxonomy of Du Châtelet's uses of the term, I believe, would be conducive for the scholarly discourse to move forward.

As Jacobs (2020, 68) states in the text quoted above, Stan (2018) is committed to the spatial reading of 'external', according to which two beings are external to each other in the sense that they are 'located in space at some distance from each other'. The first question is, is Stan indeed committed to this view? A careful reading of Stan's text shows that the answer is 'no'. Stan (2018) is a comparative project, which juxtapositions the conceptual counterparts in the metaphysics developed by the mature Leibniz, Wolff, and Du Châtelet to argue for the conclusion that contrary to the standard view, 'Du Châtelet's main source of inspiration was Christian Wolff',

---

[4] The Brading texts here referenced are Brading (2018) and Brading (2019), whereas the Stan text is Stan (2018).

not Leibniz (Stan, 2018, 477). While the scope of Stan's study is wide-ranging, nowhere in it does he appear to have an interpretation of §77. In fact, Stan did not even set foot in Du Châtelet's chapter on space. Regarding Du Châtelet's views on space, Stan only passed a fleeting remark that '[f]or her, space and time are grounded in real relations between elements, not just abstractions from common perceptions by all basic substances' (489). Jacobs might have independent reasons to believe that implicit in what Stan says is a commitment to a particular interpretation of §77, and yet since he did not cite any source text from Stan to make his reasons explicit, I believe that it is only fair not to attribute the spatial reading to Stan. Thus, insofar as the point of contention concerns Du Châtelet's use of 'external' in §77, Stan is not a contending party in the debate.

Here comes the second question: setting aside Jacobs' mistake identified above, is his framing of the debate in terms of a spatial reading vs. an ontological one of 'external' nonetheless a valid one? The answer, as I am going to show next, is again negative, for his framing had neglected the complexity surrounding Du Châtelet's uses of 'external'.

The complexity is largely owing to the fact that §77 is cited as a premise to argue for the conclusion that we must represent simple beings as spatially extended in §133. Jacobs is undoubtedly aware of this connection. However, what he hasn't noticed is another equally important feature of §133: there, Du Châtelet uses 'external' three times, each time denoting a different kind of relation. Now, let us look at §133 and distinguish Du Châtelet's three uses of 'external':

> Furthermore, it is by this dissimilarity that one can understand how non-extended Beings can form extended Beings; for the Elements exist each of them necessarily *external to the others* (since one can never be the other), and all of them being, as we have just seen, united and linked together, an assembly of several diverse Beings results from this, each of which *exists external to the others*, and which by their interconnections make a whole. But I have shown that we cannot represent extension other than as an assembly of several diverse, coexisting things, and which *exist external to one another* (§77): therefore, conclude the Leibnizians, an aggregate of simple beings must be extended. (§133, my emphasis)

Contained in the above quotation is Du Châtelet's explanation of why an aggregate of non-extended simple beings must nonetheless appear to us as extended bodies. Note that in this short passage, she uses 'external' three times, and the three uses may not be univocal. In the first occurrence, 'for the Elements exist each of them necessarily external to the others,' Du Châtelet points out something metaphysically salient about how simple beings exist independently of how we represent them to ourselves. The external-to relation at issue here is a metaphysical relation, for it is grounded in the uniqueness of internal determinations of each simple being ('since one can never be the other'). In the second occurrence, 'each of which exists external to the other,' the relata of external-to are no longer simple beings, but ordinary objects that 'result from' the interconnection of simple beings (such as tables and chairs). In the third occurrence, 'we cannot represent extension other than as an assembly of several diverse, coexisting things, and which exist external to one another,' Du Châtelet is referring back to her §77 discussion about relating objects as external to each other in the imagination. Since the relata entering the three occurrences of 'external' are different (i.e. simple beings, objects arising from simple beings, and imagined objects), committing to a particular interpretation for one use of 'external' need not commit one to that same interpretation for another. Consequently, before setting up a debate about what external-to means in Du Châtelet, we need to first specify, among the three candidates, which use of the term is at the center of debate.

In defense of Jacobs, one might object that even though 'external' is used three times to denote the relations among three kinds of relata, it does not follow that it denotes three distinct kinds of relations.[5] For example, consider dissimilarity. One could argue that for someone committed to the Principle of Sufficient Reason (e.g. Du Châtelet), this same relation would obtain among all simple substances on the ground tier as well as all bodies on the phenomenal tier; otherwise, God would have no reason to distinguish among the substances or among the bodies. Therefore, we need not separate the various uses of 'external' in this passage, at least not solely on the grounds that this term denotes relations among three different kinds of beings.

Response: This is a legitimate concern. Merely noting that 'external-to' obtains among three kinds of relata does not, by itself, establish that it does so as three distinct kinds of relation. However, the original contention is not that each use of 'external' corresponds to a distinct kind of relation. Rather, the proposal is that, since the relata in question are ontologically distinct, the interpretive considerations applicable to one are likely inapplicable to the others. Consequently, commentators would benefit from distinguishing between the different uses of 'external' before taking a position in the debate surrounding this term. Let me illustrate this with a concrete example. Du Châtelet advocates for a division of labor between the faculties of understanding and imagination, emphasizing that we cannot represent simple beings to ourselves through the latter:[6]

> It must be confessed that this conclusion astonishes the imagination, simple beings are not within its province, they cannot be represented by images, and only the understanding can conceive of them. (§120)[7]

Since simple beings are not suitable intentional objects for our imagination, it follows naturally that we cannot represent them as external to each other imagistically. Instead, as an inter-substances relation, external-to can only be thought, e.g. as grounded in the dissimilarity relation: 'for the Elements exist each of them necessarily *external to the others* (since one can never be the other)'. By contrast, in §77, where Du Châtelet discusses how we form the idea of extension by representing objects as

[5] I am grateful to Jeffrey McDonough, the co-editor of this volume, for making this objection.

[6] This division of cognitive labor received an in-depth discussion in Carus (2024), which emphasizes that for Du Châtelet, we can only know of simple beings (e.g. its existence and basic properties, such as force-bearing) through the faculty of understanding, whereby the Principle of Sufficient Reason is used.

[7] The conclusion mentioned in the first clause is one concerning the existence of simple beings and their being the ground of the existence of extended bodies. For her arguments for the existence of simple substances, see §120.

external to each other, the faculty at work is the imagination.[8] The external-to relation in §77, therefore, is a relation that *can* be represented imagistically. Given these considerations, we would do a better service to the text by separating our discussions about external-to as a metaphysical relation, which concerns how simple beings exist external to each other, and external-to as an epistemic relation, which concerns how we represent objects as external to each other via the imagination. Going forward in this chapter, we shall focus on the latter only and put Jacobs' proposed reading under scrutiny again.

4.4 Against the Ontological Reading

Having isolated the proper object of contention, we are now ready to launch a critical assessment of Jacobs' ontological reading of *external*. According to this reading, when we distinguish objects numerically via the imagination, we do so by imagining them as external to each other in the sense of representing them as ontologically independent of each other. In this section, I will engage with Jacobs' arguments for the ontological reading and show why they are lacking.

4.4.1 The Problem of Historical Precedent

The first argument Jacobs advances is historical, which appeals to Wolff's use of the term 'external' as 'a strong historical precedent' of Du Châtelet's. Jacobs writes:

> First, note that Wolff uses the term 'external' in a non-spatial sense. *Insofar as Du Châtelet follows Wolff*, then, the weak [ontological] reading is preferable. According to Wolff, it is the ontological order of externality that we represent as a spatial order. (Jacobs, 2020, 68)

While it is true that Du Châtelet references Christiaan Wolff on several occasions in the *Foundations,* I think that as commentators, we need to be very cautious when we attempt to use resources from Wolffian texts or Wolff scholarship to handle interpretive issues in Du Châtelet.[9] As I am going to show next, such caution is especially needed for the problems surrounding Du Châtelet's sense of external-to.

---

[8] 'If in our imagination we represent to ourselves a structure that we have never seen before; [...] *Likewise*, if we represent two men ideally, or even the same man twice, we place them external to one another, because we cannot force our mind to imagine that they are one, and two, at the same time' (§77, my emphasis). That the only relevant faculty in §77 is the imagination is further supported by a later passage, 'When we have thus formed a Being in our imagination from the diversity of the existence of several things and of their union, extension' (§79).

[9] Thanks to Hanns-Peter Neumann (2022), we now have a clearer understanding of the Wolffian sources to which Du Châtelet had direct access. According to Neumann, while the French translation of Wolff's *German Metaphysics* had certainly influenced Du Châtelet's the *Foundations*, the extent of this influence remains an open question. Additionally, it is intriguing that Du Châtelet explicitly references Wolff's *Ontologia* in her footnotes instead of *German Metaphysics*. I am thankful to

First, it turns out that Wolff does not deploy 'external' univocally across his writings, either.[10] As a consequence, in order to invoke Wolff as a 'strong historical precedent' of Du Châtelet, one must first overcome two enormous difficulties concerning Wolff: first, distinguishing various uses of the terminology in Wolff (like those done for Du Châtelet in the previous section) and, second, determining which one plays an important role in shaping Du Châtelet's thinking. Neither is an easy task. For instance, Jacobs cites a passage from Wolff's German text, *Deutsche Metaphysik* (1720), as well as Hogan (2007)'s interpretation of it, to support his verdict that Wolff understands external-to in terms of ontological independence, as opposed to spatial separateness. Jacobs writes:

> Consider for example the following passage from Wolff's *Deutsche Metaphysik*: §46. *What space is.* Now when many things that exist at the same time and are not identical are represented as external to one another (§45), a certain order among them thereby arises [...]. And as soon as we represent this order [of externality] to ourselves, we represent space to ourselves. (Jacobs, 2020, 68)

However, note that *Deutsche Metaphysik* isn't the Wolffian text Du Châtelet makes reference to in the *Foundations*, but the Latin text *Philosophia Prima* Sive Ontologia (1736).[11] To establish Wolff's use of 'external' as a 'strong historical precedent' to Du Châtelet's, the text she finds worthy of reference ought to be given priority to those she makes no mention of. Second, to complicate the matter further, it seems that in the *Ontologia* chapter Du Châtelet references, Wolff uses 'external' in a spatial sense. Consequently, even if we grant that Du Châtelet follows Wolff in her use of the said term, what we ought to conclude is that she, too, uses the term in a spatial sense. Let us take a closer look at the relevant passage. Wolff writes:

> If, when attending to the co-existence of simultaneous objects A, B, C, D, we distinguish the manner in which A co-exists with B from the manner in which it co-exists with the other objects C and D, and likewise we distinguish the manner in which B co-exists with C from the manner in which it co-exists with the other objects A and D, and so on; inasmuch as these objects are placed in a certain order with respect to each other, such that the *distance* between A and C be distinct from that between A and D, and so on; then we have the notion of space. The truth of this proposition is clear by experience, so long as we attend to whatever objects nearest to us. Even more, we cannot even imagine (imaginari) things as co-existing *external to each other*, without the notion of space attached to them. (Wolff, 1736: §588, *Ontologia*, my emphasis.)[12]

Clara Carus, the co-editor of this volume, and Aaron Wells, for their helpful discussion with me over the complexity surrounding Du Châtelet's intellectual relationship to Wolff.

[10] This should not be surprising, as Wolff is famous for being extremely prolific. He authors more than 50 (most multi-volume) titles, in addition to dozens of shorter essays and prefaces and nearly 500 book reviews' (Hettche & Dyck, 2019).

[11] It is noteworthy that Du Châtelet references Wolff's *Ontologia* only in the 1740 edition. More specifically, she references the following chapters: 'De Principio Contradictionis, de Principio Rationis Sufficientis, de Possibili, and Impossibili, de Necessario and Contingente, de Extensione, Continuitate, Spatio, Tempore (Of the Principle of Contradiction, of the Principle of Sufficient Reason, of Possibility, and Impossibility, of Necessity and Contingency, of Extension, Continuity, Space, Time)'. This reference appears in Section XII of the 1740 Preface but is absent in the 1742 edition.

[12] I am very thankful to Bryce Gessell for sharing his translation of this text with me.

In this passage, Wolff's goal is to give an account of 'the manner in which one arrives at the notion of space'. Thematically, this passage is highly akin to Du Châtelet's §77–§79 in the *Foundations*, which purports to explain 'how we came to form our ideas of extension, Space, and continuity' (§77). Note that Wolff uses 'external' in the penultimate clause, 'we cannot even imagine things as co-existing external to each other, without the notion of space attached to them'. The question is, in which sense does Wolff employ this term 'external' here? In light of his purpose and argumentation in this quotation, I shall argue that Wolff intends a spatial sense to externality in this particular context.

According to Wolff, we obtain the notion of space by way of attending to the manner of coexistence of objects existing simultaneously. As Wolff characterizes it, this procedure requires two mental operations: coupling and distinguishing. First, given a multiplicity of objects (e.g., A, B, C and D) that coexist, the mind couples every object with the others into pairs (e.g., A and B, A and C, A and D, B and C, B and D, C and D). Second, the mind is then to distinguish between the distance relations holding between the objects in each pair, thereby obtaining the notion of space. It is beyond the scope of this paper to critique Wolff's views here. For our purpose, notice that Wolff insists that his account holds for objects that are perceived as coexisting as well as those that are *imagined* as such: 'we cannot even imagine things as co-existing external to each other, without the notion of space attached to them'. In other words, in order to obtain the notion of space, Wolff thinks that we need not begin with a sensory perception of objects that coexist. Rather, we may also begin with similar scenarios presented to the mind by the imagination. We could arrive at the notion of space by applying the same mental operations to imagined objects as we would with objects actually perceived as coexisting. Now, on the assumption that the procedures to be followed for actually perceived objects and imagined objects are identical, when Wolff says 'imagine things as coexisting external to each other,' he should be taken to mean imagining objects as bearing distinct distance relations from others. 'External', therefore, is here operating in a spatial sense. No issues pertaining to the objects' ontological status appear to matter in this passage.

One might object that, although Wolff makes no explicit mention of ontological issues in this passage, the ontological reading is compatible with the proposed scenarios. We could begin with a multiplicity of objects perceived as external to each other in the ontological sense, i.e., as each independent of the others for its own existence ('A, B, C, D'). With objects thus given, the mind could then follow the same procedure: of coupling the objects into pairs, distinguishing the distance relations holding for each pair, thereby obtaining the notion of space.

In response, I would point out that the compatibility is only apparent. The objection construes the representation of multiple objects as each existing independently of the other as the given, which is then subject to two further mental operations of coupling and distinguishing. In other words, to perceive objects as external to each other (in the sense of existing independently of each other) is not yet sufficient for the notion of space to arise. However, notice Wolff's wording, which suggests something quite different: 'we cannot even imagine things as co-existing external to each other, without the notion of space attached to them'. Here in this remark, the representation of objects as coexisting external to each other is said to *suffice* for the notion of space to arise: no further mental operations are needed. Hence, the ontological reading of 'external' ought to be rejected for this text.

In conclusion, even if Wolff's *Ontologia* plays an important role for Du Châtelet's thinking about external-to in the context of a similar philosophical problem (i.e. how do we obtain the notion of space?), as we have seen, the relevant passage in the *Ontologia* does not support an ontological reading. With respect to external-to, therefore, we ought to resist using Wolff as a 'strong historical precedent' to interpret Du Châtelet.

4.4.2 The Problem of Mismatch

Now, let us focus on a different part of the same text ('we cannot represent to ourselves several different things as being one, without this resulting in a notion [...] we call *Extension*') and see why the ontological reading fails again. According to the ontological reading of 'external', we represent numerically distinct things as ontologically independent of each other. But how might things that are represented as ontologically independent of each other also appear to us 'as being one'? Noting that Du Châtelet has not explicitly told us what it is to represent objects 'as being one,' Jacobs writes:

> The sub-conclusion (3) states that when we represent a multiplicity as a unity, we necessarily represent it as spatially extended. However, we don't yet know if the antecedent of this conditional is true: do we 'represent to ourselves several different things as being one'? While it is clear that Du Châtelet assumes this, she does not discuss this assumption in Chap. 5. (Jacobs, 2020, 69)

The 'sub-conclusion (3)' mentioned in this quotation comes from Jacobs' reconstruction of §77 based on his proposed ontological reading of 'external.' Let us examine this sub-conclusion in light of the two premises it is supposed to follow from:

(1) We necessarily represent numerically distinct simples (i.e. multiplicities) as external to each other.
(2) We represent simples that are external to each other as a unity by representing them as spatially extended.
(3) Sub-Conclusion: When we represent a multiplicity as a unity, we necessarily represent it as spatially extended. (From (1) and (2))

In Jacobs' reconstruction, what is being represented as external to each other are 'simples,' namely, simple substances on the ground floor. This maneuver is problematic, as it applies what §77 says to simple beings directly, thereby overlooking the division of labor that Du Châtelet establishes between the faculties of understanding and imagination. Recall that in Sect. 4.1, we discussed this division as imposing a strict constraint on the interpretation of the text: on one hand, §77 states that we obtain the idea of extension by imagining 'several different objects as being one';

on the other hand, Du Châtelet asserts that simple beings are not suitable for presentation through the imagination: 'Thus, simple beings are completely different from compound beings, and they cannot be seen, touched, or represented in the imagination by any perceivable image'(§123). There is, as it were, a glaring mismatch between the target passage, §77, and the type of intentional objects broached in Jacobs' reconstruction, namely, simple beings.

Now, one might wonder what it means to represent simple substances as external to each other via the understanding, as opposed to the imagination. Although the text does not provide a direct answer to this question, we can glean an educated guess from the following passage: "the Elements exist each of them necessarily external to the others (since one can never be the other)' (§133). Here, representing simple substances as external to one another through the understanding appears to require that we conceive of externality, as an inter-substance relation, as grounded in the non-identity of simple substances. In doing so, we can infer one property that can be predicated of simple substances (mutual externality) from another (non-identity). We may derive additional predicates through metaphysical reasoning. However, as long as the faculty at work is the understanding, we could not arrive at the idea of extension through the representation of the external-to relation in the manner described in §77, since Du Châtelet explicitly states there that it is the imagination that performs this feat.

While Jacobs' reconstruction ultimately fails, his question remains pertinent: in what sense do we represent several different things 'as being one'? In the article, Jacobs references a passage from Chap. 7 to address this problem. However, I believe the answer can be found right here in Chap. 5. Consider these passages: In §79, Du Châtelet states, 'Thus, as long as we consider the possibility that several different things can *exist together* in this abstract Being we call *extension*, we form the notion of Space' (§79, first emphasis mine); similarly, in §80, she remarks, 'For, insofar as we envisage *several things existing together*, and stripped of all internal determinations, this Being appears to us to be enduring,' where 'this Being' refers to extension as perceived by the mind (§80, emphasis mine). By comparing the content of these passages, we can confidently infer that the phrase 'several different things exist[ing] together' and 'several things existing together' in §79 and §80 are synonymous with 'several different things as being *one*' and 'this diversity and this union' in §77. In other words, 'existing together' is synonymous with 'being one' and 'this union.' Objection: The proposed reading of 'union' in terms of togetherness is *not* incompatible with Jacobs' ontological reading of 'external'. In fact, once we substitute simple beings in Jacobs' reconstruction with ordinary objects and apply the appropriate faculty—the imagination—there is nothing inherently problematic about the ontological reading itself. Response: The ontological reading of 'external' does not align naturally with the text, as it is challenging to see how a representation of mutually independent objects constitutes a representation of these objects as existing *together*. How would the spatial reading address this challenge? Let us consider an apple and a pear as spatially external to each other. By imposing a spatial relation to the objects, our minds form a representation of their existing together or, coexistence, because in virtue of standing in spatial relation with each other, the objects are co-spatial and, as such, they appear to us as co-existing. The spatial reading, therefore, provides a much better accommodation to the text.

4.4.3 The Problem of Extended Souls

The last argument against the ontological reading we shall examine draws on a different passage, also in §77. Du Châtelet writes:

> *It is indeed true that diversity and union engender in us the idea of extension*, and that several Philosophers have wanted to treat our soul as something extended because they noticed in it several different faculties that nevertheless constitute a single subject. In this they were mistaken: to regard attributes and modes of a Being as separate Beings, *existing external to one another*, is to abuse the notion of *Extension*; for these attributes and modes are inseparable from the Being that they modify. (§77, my emphasis)

In this quotation, Du Châtelet contends that it is a mistake to attribute extension to the soul even though the soul is some sort of unity containing diversity (or, as she puts it, 'a single subject with several different faculties'). According to Jacobs, the reasoning in this passage requires an ontological reading of 'external'. He writes:

> Du Châtelet argues that, since the attributes and modes of a Being are inseparable from it, it is incoherent to consider them as external to each other. It is again implausible that she is concerned with the physical location of these attributes and modes. Rather, it seems that the sense of internal and external used here is again that in which attributes are internal to the substance in which they inhere, whereas distinct substances are external to each other and thus ontologically independent. Du Châtelet's point then is that, since attributes and modes are internal to and dependent on Simple Beings (as she argues in Chap. 3), it is a mistake to represent them as external to each other. (Jacobs, 2020, 68)

Jacobs reads Du Châtelet's reasoning in this passage as something like the following:

(1) Some philosophers hold that the soul is extended because we represent the soul's diverse attributes and modes as external to each other, that is, as ontologically independent existents.
(2) The attributes and modes of a Being cannot exist independently of the Being in which they inhere.
(3) Therefore, it is a mistake to represent attributes and modes as ontologically independent from each other.
(4) Moreover, it is a mistake to hold that the diverse attributes and modes of the soul constitute an extended being.

In my view, however, Jacobs' reading is misguided because it fails to make sense of an important sentence in the beginning of the passage: 'it is indeed true that diversity and union engender in us the idea of extension'. On Jacobs' reading, those who hold that the soul is extended are just bad at metaphysics: they fail to understand the ontological dependence relations among different metaphysical kinds. In my mind, however, Du Châtelet's point is significantly subtler than that. Du Châtelet aims to defeat the mistake by explaining *why* it is such a tempting mistake to make. Her explanation is that when the imagination represents diversity and union, it always does so by presenting those diverse elements as constituting a unified, extended thing. Any good interpretation of 'external,' therefore, must make good Du Châtelet's claim that when we represent several different things as constituting a unity by placing them external to each other in the imagination, we represent them as forming an extended thing. That is to say, by Du Châtelet's light, this feature of the mind explains why some philosophers tend to attribute extension to the soul.

On Jacobs' proposed ontological reading, however, this claim ('it is indeed true that diversity and union engender in us the idea of extension') does not hold. For we *can* indeed conceive of different cognitive faculties (diversity) as ontologically independent of each other while constituting the same subject (unity) without that conception engendering the notion of extension. For example, consider the faculty of understanding and that of imagination. Both faculties are modes of the soul and therefore dependent on the latter for their existence. In representing the understanding and the imagination as ontologically independent of to each other, we might, for instance, conceive of them as two faculties of the soul which are distinguished by the intentional objects of their respective powers (e.g., only the understanding can present simple substances to us, whereas the imagination cannot, as Du Châtelet points out in §127). Nothing in this representation seems pertinent to the idea of extension at all, which would make it strange for Du Châtelet to claim that 'it is indeed true that diversity and union engender in us the idea of extension.' The ontological reading of 'external,' therefore, fails to accommodate Du Châtelet's point.

One may object on Jacobs' behalf that Du Châtelet does *not* intend the claim that 'diversity and union engender in us the idea of extension' to be a true statement about what the representation of diversity and union engenders. Indeed, she ultimately shows that a mental representation of several distinct faculties of the soul as constituting a unity cannot yield the idea of an extended soul. Therefore, a good interpretation need not accommodate the claim as it stands.

To this I respond that the objection is illegitimate. Close inspection reveals three layers to the reasoning in this passage, and a good interpretation must preserve all of them. First, Du Châtelet acknowledges that the idea of extension seems to have some intimate connection with our representations of diversity and union. Second, she then immediately cites this intimate connection as the *explanans* for why 'several Philosophers have wanted to treat our soul as something extended' (§77). Finally, she offers the correct metaphysics of the soul. The first two steps of this chain of reasoning are closely linked in that the first explains the second Therefore, a good interpretation of 'external' should not only make it true that our representation of diversity and union engender extension, but this fact must also be explanatory of the mistake made by extended-soul theorists.

A spatial reading of 'external' does an excellent job at accounting for these levels of reasoning. Why is it 'indeed true' that 'diversity and union engender in us the idea of extension'? The first thing to be noted in seeking an answer to this question is that all of Du Châtelet's examples in §77 involve the imagination only. Hence, her §77 discussion of diversity and union should be taken as referring to the representation of diversity and union via the imagination. Thus, Du Châtelet's point is that when one uses the imagination to represent distinct faculties of the soul as modes inhering in it, one inevitably obtains a representation of the soul as an extended thing. Consider how we ordinarily represent the soul to ourselves via the imagination. An introductory course in Philosophy of Mind might have its students draw a circle on the board to represent the soul and then divide it into several compartments, each standing for a distinct cognitive faculty, such as memory, imagination, and will. The image-based representation serves a great pedagogical purpose, representing those faculties as distinct mental kinds that still belong together in a thing ordinarily referred to as 'the soul.'

Now, such a diagram represents faculties of the soul as *spatially* external compartments constituting a single *spatially* extended entity, but if a student takes the diagram to be an accurate representation of the soul, i.e. that the soul really is an extended entity constituted by several spatially external components, she is mistaken. At this point, Du Châtelet's second step in her reasoning becomes relevant. She suggests that it would be a mistake to take the diagram to reveal the nature of the soul. That is, it would be a mistake to think that those faculties are really *spatially* external to each other in just the way that their representative compartments are. It would likewise be a mistake to conclude that the soul to which the faculties belong is really *spatially* extended like the circle representing it. Those rushing to such judgments are liable to do so because they are too trusting of what the imagination presents to them. It turns out that the imagination frequently paints a false picture about the nature of things. For instance, one might believe that space is empty because one can easily imagine the complete absence of bodies by abstracting from their internal determinations (e.g. shapes and colors), thereby making it seem as if 'there is nothing left in this Space' (§82). But is space really empty? To answer this question, Châtelet would encourage us to draw on metaphysical resources. In an earlier passage, she presents an argument against empty space by appealing to the Principle of Sufficient Reason. According to this argument, space must be plenist because the shapes of bodies require explanation, which can only be found in the surrounding matter.[13] This is the same diagnosis that Du Châtelet issues to those extended-soul theorists who arrived at their false beliefs by trusting in the imagination's representation of the soul as a unity with distinct parts.

A further question the student may have is, how could we ever learn about the nature of the soul, if we cannot trust the spatially extended image-based representation found in our imagination? This question is perhaps addressed in the last layer of Du Châtelet's reasoning. Instead of trusting the imagination, we should engage in serious metaphysical inquiry in order to discover the soul's true nature. Such an inquiry is the main work of Chap. 3 in the *Foundations*, in which Du Châtelet distinguishes between essence, attributes, modes, and their respective relation to the subject in which they inhere. That investigation reveals that the existents in these

[13] This argument was recently noted and discussed in McDonough (*forthcoming*): "Émilie Du Châtelet Against Spatial Metaphysics" in *The Bloomsbury Handbook of Du Châtelet*, ed. Fatema Amijee.

categories are related by relations of ontological dependence as opposed to spatiality. For example, modes and attributes depend on the substance in which they inhere for existence, while modes do not depend on attributes for existence. Only when we see the fittingness of invoking relations of ontological dependence to conceive of the soul's relation to its diverse faculties can we be immune from the mistake of taking the compartmented circle as revealing the soul's true nature.

Finally, not only can the spatial interpretation of 'external' advanced in this paper adequately accommodate all three steps in Du Châtelet's reasoning, but it possesses the further virtue of aligning with the overarching thrust of Du Châtelet's theoretical work on the idea of space in Chap. 5. In this chapter, Du Châtelet takes the absolute-relative debate on the nature of space as her point of departure and seeks instead to provide an epistemology of the idea of space. Metaphysically, she rejects absolute space as 'an illusion of the imagination' and sides with relationalists in thinking of space as the order of relation of coexisting objects. Epistemologically, however, she is concerned with explaining why we *must* represent space as *absolute*; that is, as an homogeneous, empty, penetrable, and immutable entity existing independently of the material bodies within it. While a full analysis of her account cannot be supplied in this paper, it suffices to observe her characteristic strategy. She contributes to a debate in metaphysics by (1) providing an epistemology for the central ideas involved, and (2) using that epistemology to explain the attraction of her opponents' view. The §77 discussion of the soul is likewise in line with this strategy. She addresses the metaphysical controversy over extended souls by first acknowledging the attraction of this idea, then locating the reason for that attraction in the ways in which the imagination presents the soul to us, before finally showing that the image-based representation cannot reveal the soul's true nature.

4.5 Conclusion

In this chapter, I revisited the scholarly debate surrounding 'external' in Du Châtelet's chapter 'Of Space' in the *Foundations of Physics* and examined the two competing readings of this term in the literature: the ontological reading and the spatial reading. After raising three problems associated with the ontological reading and showing that the spatial reading does a better job at accommodating the relevant text, I argue that we should favor the latter interpretation over the former.

References

Brading, K. (2018). Emilie Du Châtelet and the problem of bodies. In E. Thomas (Ed.), *Early modern women on metaphysics* (1st ed., pp. 150–168). Cambridge University Press.

Brading, K. (2019). *Émilie Du Châtelet and the foundations of physical science*. Routledge.

Carus, C. (2024). Émilie Du Châtelet's metaphysics in light of her concept of a being. *Journal of Modern Philosophy* 6.

Du Châtelet, E. (1740). *Institutions de Physique*. Paris. Translated by Brading et al. (2018). http://kbrading.org/translations

Hagengruber, R. (2011). Du Châtelet and the transformation of metaphysics. In Hagengruber (Ed.), *Du Châtelet between Leibniz and Newton* (pp. 1–59). Springer.

Hettche, M., & Dyck, C. (2019). Christian Wolff. In E. N. Zalta (Ed.), *The Stanford encyclopedia of philosophy* (Winter 2019 ed.). https://plato.stanford.edu/archives/win2019/entries/wolff-christian/

Hogan, D. (2007). Wolff on order and space. In J. Stolzenberg (Ed.), *Wolff und die Europäische Aufklärung: Akten des 1. Internationalen Wolff-Kongresses. Hildesheim.* Georg Olms Verlag.

Hutton, S. (2012). Between Newton and Leibniz: Emilie du Châtelet and Samuel Clarke. In R. Hagengruber (Ed.), *Emilie du Châtelet between Leibniz and Newton*. Springer.

Jacobs, C. (2020). Du Chatelet: Idealist about extension, bodies and space. *Studies in History and Philosophy of Science Part A, 82*, 66–74.

McDonough, J. (forthcoming). Émilie Du Châtelet against spatial metaphysics. In F. Amijee (Ed.), *The Bloomsbury handbook of Du Châtelet*. Bloomsbury.

Neumann, H. P. (2022). Émilie Du Châtelet within the correspondence between Christian Wolff and Ernst Christoph of Manteuffel. In R. E. Hagengruber (Ed.), Époque Émilienne. *Women in the history of philosophy and sciences*, vol 11. Springer, Cham. https://doi.org/10.1007/978-3-030-89921-9_7

Reichenberger, A. (2021). Émilie Du Châtelet on space and time. *Revue D'histoire des Sciences, 2021*(2), 331–355.

Stan, M. (2018). Emilie du Châtelet's metaphysics of substance. *Journal of the History of Philosophy, 56*(3), 477–496.

Wolff, C. (1720). *Vernünfftige Gedancken von Gott, der Welt und der Seele des Menschen, auch allen Dingen überhaupt* (*Deutsche Metaphysik*). Frankfurt.

Wolff, C. (1736). *Philosophia prima, sive Ontologia*, editio nova. Frankfurt, [*Ontologia*].

Chapter 5
Extension and Space in Chapter V of *Institutions de physique*

Katherine Dunlop

At the opening of Chapter V of *Institutions de physique*, "On Space", Du Châtelet signals the importance of the metaphysics of space. She characterizes the question of the nature of space as one of the most famous in philosophy and one of the most essential for "Physics and Metaphysics" (V.72; 1740/90, 1742/94). Without denying the importance this metaphysical issue has for Du Châtelet, in this chapter we will look beyond her treatment of it to her account of spatial representation, in order to thereby bring out the originality of her contribution.

Following Leibniz's lead (in his correspondence with Clarke), Du Châtelet frames her positive contribution to the theory of space as an account of how we form the idea of space. But while this appears to be mere framing on Leibniz's part (at least in the correspondence), it is a central goal of Du Châtelet's to explain how the idea of space arises, and specifically to show that it is formed by adjoining further content to the highly abstract idea of extension.[1] My aim in the first half of this chapter is to introduce Du Châtelet's project by contrasting it with other approaches. In Sect. 5.1, I build on Jeffrey McDonough's case for rejecting the prevailing interpretation of Chapter V as a defense of Leibnizian relationalism. McDonough argues that Du Châtelet advocates a return to the "spaceless metaphysics" of Descartes and Aristotle. I aim to complement McDonough's reading by showing how Du Châtelet can also be seen to engage in a more forward-looking epistemic project. Section 5.2 brings out some respects in which Du Châtelet follows Christian Wolff more closely than Leibniz. Most notably, for both Wolff and Du Châtelet the idea of extension, which represents

[1] This paper was first written independently of Qiu Lin's (*ms.*), which also emphasizes the thinness of the idea of extension as compared with the idea of space. I am very grateful to Lin for making her manuscript available and for conversation, both of which have led to significant improvements.

K. Dunlop (✉)
Department of Philosophy, University of Texas at Austin, 2210 Speedway, Austin, TX 78712, USA
e-mail: kdunlop@utexas.edu

C. Carus and J. K. McDonough (eds.), *Émilie Du Châtelet in Relation to Leibniz and Wolff*, Women in the History of Philosophy and Sciences 23,
https://doi.org/10.1007/978-3-031-80541-7_5

things that are "external to" each other as united together, is prior to the idea of space. Section 5.3 takes up the issue of how to understand the term "external" in Du Châtelet's account of extension. In the literature, two readings of "external"—as spatial removal, or ontological distinctness—have been distinguished. I defend a spatial reading, as more congenial to the understanding of Du Châtelet's project as epistemic. But if we give "external" a spatial meaning and understand the project as epistemic, her account of space appears circular. So it is necessary to explain (Sect. 5.3.1) how the representation of particular instances of spatial relationships can be the basis for the representation of space (as a homogeneous, unbounded, encompassing whole). Circularity threatens once again if the basis for the representation of space is supposed to be empirical, as on Wolff's account. In Sect. 5.3.2, I argue that Du Châtelet may escape this objection because her concern is just to explain how the idea of space is related to that of extension, not to find an empirical origin for either.

Although Wolff's and Du Châtelet's accounts of extension are similar, and for both the idea of extension is prior to that of space, Du Châtelet offers an original view of the relationship between the ideas of extension and space. Section 5.4 compares the ideas of extension and of space in order to show that on her account, the idea of extension lacks the content that would be needed to determine its ontological status. On this interpretation of the distinction between extension and space, only space, not extension, can be represented—or as Du Châtelet says, "appear to us"—as having any particular ontological status. So it presents a difficulty for my reading that Du Châtelet says that extension appears to us as a substance. In Sect. 5.4.3, I argue that in these contexts she is working with a different notion of extension.

The final sections of the paper discuss how and why Du Châtelet expounds the capacity through which we represent extension. Section 5.5 considers Du Châtelet's use of the example of a line to illustrate how we represent extension. This is another juncture at which the materials Du Châtelet provides to help us understand (the representation of) extension seem to already involve what she seeks to explain. But, I argue, the choice of this example is apt for illustrating what she describes as the power to "form, by abstraction, imaginary beings" (V.86). One reason for Du Châtelet to discuss this capacity is that she argues, following Wolff, that the Newtonian conception of absolute space is merely "imaginary".[2] Section 5.6 identifies a further reason, more integral to *Institutions*: to bring into a view a positive role for abstractions in science, which was obscured by her earlier strictures on the misuse of hypotheses.

[2] Du Châtelet's "error theory" is discussed by Lin (*ms.*) and McDonough (*ms.*). I discuss Wolff's treatment of absolute space in Dunlop 2013.

5.1 Relationalism in Chapter V of *Institutions*

In his exchanges with Clarke, Leibniz is clearly contending for a view of space as not only relational, but ideal. That his account of the formation of the idea of space is in the service of his idealist spatial metaphysics is apparent from the overall context, and confirmed by the passage's conclusion. On Leibniz's account, we "observe in" a number of coexisting things "a certain order of coexistence"; then come to represent place in terms of that order; and, finally, represent space by representing the places that it "comprehends". In conclusion, Leibniz claims to have explained "why men fancy places … and spaces; though these things consist only in the truth of relations, and not at all in any absolute reality" (V.47).

It is not obvious, to me at least, why Leibniz gives his account of the idea's formation at precisely this juncture in the correspondence. But his point may be to clarify the metaphysical status of space by showing how discourse about space can be interpreted in terms of what he elsewhere calls "*concreta*". Leibniz holds, in particular, that we come to represent places by applying a criterion for sameness of place. He first explains place in terms of sameness of relation:

> When it happens that one of these co-existent things changes its relation to a multitude of others, which do not change their relation among themselves; and that another thing, newly come, acquires the same relation to the others, as the former had; we then say, it is come into the place of the former … (V.47)

Leibniz then explicitly defines sameness of place:

> And supposing, or feigning, that among those co-existents, there is a sufficient number of them, which have undergone no change; then we may say, that those which have such a relation to those fixed existents, as others had to them before, have now the *same place* which those others had. (V.47)

Leibniz draws attention to his way of proceeding in his conclusion, which begins, "To conclude: I have done here much like Euclid, who not being able to make his readers well understand what *ratio* is absolutely in the sense of geometricians; defines what are the *same ratios*. Thus, in like manner, in order to explain what *place* is, I have been content to define what is the *same place*" (V.47). He indicates elsewhere that this style of definition is appropriate for "*abstracta*" because they lack reality.

> I see no other way of avoiding these difficulties than by considering *abstracta* not as real things [*res*] but as abbreviated ways of talking [*compendia loquendi*]—so that when I use the name *heat* it is not required that I should be making mention of some vague subject, but rather that I should be saying that something is hot … Geometricians, too, do not use definitions of *abstracta* but reduce them to *concreta*: thus Euclid does not use his own definition of *ratio* but rather that in which he states when two quantities are said to have the same, greater, or lesser ratio.[3]

This analogy suggests that Leibniz is "content to define what is the *same place*" in terms of relations among "fixed existents" because the latter serve as the subjects to which spatial discourse refers.

---

[3] "De realitate accidentium", quoted in Arthur 2021, 348.

As mentioned above, Du Châtelet opens her chapter on space by characterizing "the question of the nature of Space" as highly important for physics and metaphysics. Du Châtelet proceeds to frame this question as whether space is "absolute, real and distinct from" bodies,[4] or whether it is "ideal" and "nothing over and above things". The explicit aim of *Institutions* is to present Newton's and Leibniz's views to a French audience (*Avant-Propos*). This makes it natural to expect that the aim of this Chapter as a whole is to adjudicate between Newton's conception of space as absolute and Leibniz's conception of it as ideal. Indeed, Du Châtelet proceeds to give Leibniz's arguments against absolute space, which she pronounces "irrefutable" (V.74; 1740/94, 1742/99), and which are based on the Principle of Sufficient Reason, which she describes as a "compass capable of leading us on the shifting sands of metaphysics" (*Avant-Propos* XII; 1740/13, 1742/13). So it makes sense that commentators typically understand her definition of space as the "order of coexisting things", and her account of how we form its idea, as a defense of Leibniz's view.[5]

Yet on closer examination, this opening paragraph exhibits some puzzling features when viewed through the lens of the relationalism/substantivalism dichotomy, as Jeffrey McDonough shows (*ms.*). In the first sentence, Du Châtelet "characterizes the distinction she is tracking as 'one of the most famous' questions concerning space, [which] has divided ancient and modern Philosophers alike" (*ms.* p. 5, quoting V.72).[6] As McDonough observes, Leibniz's relational theory was first propounded by Leibniz himself, so the opposition between it and Newton's substantivalism can hardly be said to have divided "ancient and modern Philosophers alike". A second oddity is Du Châtelet's claim that on the realist view, space "receives the Bodies that are placed in it" as an "infinitely extended fluid" in which they "swim". In light of Newton's own austere characterization of the parts of space as essentially places,[7] this picturesque account of the collocation of bodies with parts of space appears misleading, or at least superfluous. Thirdly, Du Châtelet states the main point of V.72, in her marginal summary, as that "Half of the philosophers believed, and still believe, in empty space and the others believe it to be filled with matter". Since the issue of whether all space is filled with matter is separate from the relationalism/substantivalism dispute, this suggests that Du Châtelet may "fixate on a consideration that is orthogonal to" the latter, as well as misrepresenting its "historical scope" and misleadingly formulating the Newtonian side (McDonough *ms.*, p. 7).

---

[4] Or, more precisely, from matter. Shortly afterwards, Du Châtelet introduces John Keill's view of absolute space, according to which "all matter is strewn with … interstices that are absolutely void, and … there is in Bodies a great deal more void than there is solid matter" (V.73; 1740/91, 1742/95). This manner of phrasing brings out that on the Newtonian view, at least in Keill's influential formulation, bodies are composites of void and matter. So the operative distinction is between void and matter, rather than matter and body.

[5] See, for example, Brading and Lin 2023, Hutton 2012, and Suisky 2012.

[6] Similarly, in V.73 Du Châtelet attributes "the view that Space is distinct from matter" to Epicurus, Democritus, and Leucippus; 1740/91, 1742/95.

[7] "Let the parts of space move from their places, and they will move (so to speak) from themselves… It is of the essence of spaces to be places" (1999, 410).

As McDonough explains, we can "make much better sense" of Du Châtelet's discussion by relating it to the dichotomy between "spaceless" and "spatial" metaphysics, instead of that between substantivalism and relationalism. "Spaceless metaphysics" and "spatial metaphysics", or "spatialism", are McDonough's terms for the exclusion or admission, respectively, of space as existing distinct from matter. Since this dispute concerns whether space (as distinct from matter) should be admitted to ontology, not the kind of being it has, Newton and Leibniz belong together on the side of spatial metaphysics; the most prominent representatives of spaceless metaphysics are Aristotle and Descartes. These views are indeed represented in ancient philosophy, by the atomists and Aristotle, respectively; and the former "did indeed sometimes suggest that space is like a fluid ... in which bodies might be said to 'swim'" (McDonough *ms.*, p. 6, citing Sedley, 1982). And Du Châtelet's focus on the existence of void space is obviously much more to the point.

The main conclusion I wish to draw is that, with McDonough, we should take Du Châtelet's arguments to target "not specifically Newtonian substantivalism but rather spatialism more generally" (*ms.*, p. 7). I also regard it as an important insight that Du Châtelet is engaging Descartes's view of extension as well as Newton's and Leibniz's views of space; on my view, she targets Descartes's view of extension (as exhausting the essence of body) in Chapter VII as well (as I argue in Sect. 5.4.3 below). To motivate a different reading of Du Châtelet's chapter on space, I just want to register a worry about how sharply the spaceless/spatial metaphysics dispute could be separated, within her own historical context, from the relationalist/substantivalist one. That is not, of course, to deny the value of retrospectively distinguishing them.

I question how sharply we can separate, specifically, Newton's positions on these issues. McDonough remarks that Newton's substantivalism is logically consistent with the world's being full of matter, and Leibniz's relationalism is likewise logically consistent with the existence of void. So when Du Châtelet appeals to the Principle of Sufficient Reason to argue against "the hypothesis of the void" (V.73–74), she appears to use "the biggest guns that Leibniz has at his disposal to argue for a view that is perfectly consistent with Newton's substantivalism" (*ms.*, p. 7). McDonough points out, in addition, that Newton defends "the postulation of the void on explanatory grounds" (*ms.*, 3), where the phenomena to be explained include Boyle's vacuum experiments as well as the observed motions of planets and comets. The empirical basis of this view can make it seem all the more distant from Newton's espousal of absolute space, particularly if we take Newton to confront the same metaphysical issues as Patrizi, Gassendi, and More (with whom McDonough aligns Newton; *ibid.*).

On one prominent line of interpretation, however, Newton's substantivalism derives its strongest support from the importance of determining true motion.[8] It is of course correct that the issue of how states of motion are determined is separate from that of the existence of void. So even if we thus connect Newton's dynamics with his substantivalism, they remain logically independent from his acceptance of

[8] The interpretation of Newton's distinction between absolute and relative space (and time, place, and motion) as expressing only what his dynamics requires has many defenders, but remains controversial. For discussion and a contrasting view, see Janiak 2012.

empty space. Yet there is a danger of underestimating the extent to which Newton's postulation of the void stands or falls with his theory of motion. McDonough quotes Newton's argument, in the Queries to the *Opticks*, that if "the Heavens" were "absolutely dense, or full of matter without any Vacuum, let the Matter be never so subtil and fluid", their resistance would greatly retard celestial motion; so "to make way for the regular and lasting Motions of the Planets and Comets, it's necessary to empty the Heavens of all Matter" (1952, 368). By itself, this does not make clear that the "regular motions" in question are motions as explained by Newton's laws (whose application Newton takes to require his distinction between absolute and relative motion). But Newton says as much in a letter to Leibniz of 1693, which finds echoes in the "General Scholium" added to the second (1713) edition of the *Principia.*[9]

> For since celestial motions are more regular than if they arose from vortices and observe other laws, so much so that vortices contribute not to the regulation but to the disturbance of the motions of the planets and comets; and since all phenomena of the heavens and the sea follow precisely, as far as I am aware, from nothing but gravity acting in accordance with the laws described by me; ... I have myself concluded that all other causes are to be rejected and that the heavens are to be stripped as far as may be of all matter, lest the motions of planets and comets be hindered and rendered irregular. (Newton, 2004, 108)

To be sure, Newton then claims he will "be far from objecting" if someone introduces "some subtle matter", "shows that the motion of the planets and the comets will not be disturbed by this matter", and "explains gravity along with all of its laws by the action of" the matter (*ibid.*). So Newton's approach to motion can be reconciled with the plenum theory to just the extent that the latter allows for the explanation of motion by "gravity acting in accordance with [Newton's] laws", which is not a matter of mere logical consistency.

Similarly, Leibniz's position on the existence of void could be taken to rest on the same basis as his relationalism. The single most important textual source for his dispute with Newton was, of course, the Leibniz-Clarke correspondence. Leibniz initiated the correspondence by cautioning Caroline, Princess of Wales, about the rise of materialism and decay of natural religion in England. As we come to recognize the importance of Caroline as the intended reader,[10] we can also appreciate how each side sought to appear more aligned with her religious views. For his part, Leibniz pushes his "single principle" of sufficient reason as what allows us to "demonstrate the being of God and all the other parts of metaphysics and natural theology" (Second Letter to Clarke, §1), taking care to show how this principle rules out the existence of space as "a substance or at least an absolute being" (Third Letter, §5) as well as the existence of vacuum (Fourth Letter, §46, Postscript) and the finitude of matter (Fourth Letter, §21).

[9] The General Scholium begins by stating that "the hypothesis of vortices is beset with many difficulties". Newton asserts that the observed rotations of the sun and major planets around their own axes, which (he adds in the third edition) "would have to agree with the motions of their vortices", differ from the periodic times that would be required for the planets to orbit in accordance with Kepler's areal and harmonic laws (1999, 939). Newton also repeats the claim to have "explained the phenomena of the heavens and of our sea by the force of gravity" (1999, 943).

[10] Caroline's importance is brought to light in Bertoloni Meli (1999) and Brown (2004).

5.2 Du Châtelet's (and Wolff's) Two-Step Account of Extension

To sum up, an eighteenth-century reader could easily come away from the Leibniz-Clarke correspondence with the impression that the question of the existence of void and the relationalist/substantivalist dispute are to be decided by the same metaphysical principles. It would then be easy to regard them as facets of one underlying controversy rather than as separate debates. The difficulty of separating out a distinct metaphysical dispute, more closely targeted by Du Châtelet's arguments, in which she could see herself to engage opens up space for other readings. My approach will be to understand Du Châtelet's treatment of the idea of space as mainly concerned with our representational capacities, rather than an account of space's manner of being.

Some obvious differences between Du Châtelet's account of the formation of this idea and Leibniz's are that she proposes to explain how "we" come [*nous venons*; V.77; 1740/96, 1742,101] to form it, rather than how "men" do [*les hommes viennent*; Fifth Letter to Clarke, §47]; and, more importantly, she considers the idea of space in conjunction with those of extension and continuity (V.77). Du Châtelet draws important connections between these ideas, which give her reason to proceed this way. She holds that we are constrained to represent space as continuous, which is important for our knowledge of physical laws, and that the idea of extension is prior to that of space.

In drawing these connections, Du Châtelet is to some extent following Christian Wolff's *Ontologia* (the only work of Wolff's that she explicitly cites in *Institutions*[11]). More generally, Wolff's treatment of the idea of space is a more likely model than Leibniz's for Du Châtelet's (understood as an examination of our representational capacities). That is because Leibniz aims to establish a metaphysical, rather than epistemic, conclusion, while Wolff—who seeks to avoid basing his views on the assumption of innate ideas[12]–is obliged to address the (epistemic) question of how we could acquire the idea of space through experience.[13]

Both Du Châtelet and Wolff precede their discussions of space with an account of how we represent extension. The account has two steps, and since Du Châtelet closely follows Wolff at each step,[14] I will focus on her version. Her first step is to argue that we must represent things that we take to be different from one another as

[11] Cf. Lin (*ms.*), n. 10.

[12] I base this claim on Wolff's treatment of concept formation in his German *Logic*. There, Wolff claims explicitly that it is not necessary to determine "whether our concepts of external things are conveyed into the soul, as into an empty receptacle, or whether ... they [lie] in the essence of the soul, and are brought forth barely by its own power, on the occasion of changes produced in our bodies by external objects" (I.6, quoted in the anonymous eighteenth-century translation printed as Abt. III V. 77 of Wolff's *Gesammelte Werke* (Hildesheim: Olms, 1962–).

[13] The literature on Wolff's discussion of space is sparse, but Brandt 2012 reads him as deriving the notion of space from experience.

[14] At *Ont.* §544, Wolff claims that we represent those things as external to each other "which we perceive as distinct from each other". At §548, he claims that the notion of extension arises when

"external to" [*hors de*] one another, even if the things exist only in thought. Thus, "if we represent ideally two men, or we repeat in our mind the representation of the same man twice, we place them externally to one another" (V.77; 1740/97, 1742/101). The second step specifies how the representation of things as external to one another gives rise, in turn, to the representation of extension. We represent extension when we "represent to ourselves several different things as being one", as for instance in a line "we pay attention to several distinct parts which we see as existing externally to one another, which are united together and which are for this reason a single whole" (*ibid.*). Here, Du Châtelet glosses "extension" as the "notion that is attached to this diversity and union"; elsewhere, she says the notion of extension "contains nothing but parts coexisting and united" (IX.169; 1740/181, 1742/190). I think we can understand "coexisting" as "existing external to one another"'.

Du Châtelet also follows Wolff in holding that the way in which we form the idea of space constrains us to represent it as continuous (*Inst.* V.81; *Ont.* §592), uniform (*Inst.* V.85; *Ont.* §598), penetrable (*Inst.* V.82; *Ont.* §599), and so forth. I maintain, however, that the respects in which she diverges from Wolff are more important than these similarities. I do not find in *Ontologia* the crucial claim that the idea of space is made up of the idea of extension and a further component (namely, the representation of "coexisting beings" as possibly bearing individuating determinations). Furthermore, Du Châtelet is not committed to finding an empirical origin for the idea of space.[15]

5.3 Interpreting Du Chatelet's Term 'External'

To understand how we represent extension, on Du Châtelet's account, we must know what she means by "external". Two construals have been sharply distinguished by Caspar Jacobs. On what Jacobs calls the "stronger" reading, the notion of externality already involves spatiality, so that we represent two things as external to one another by representing them as spatially removed from one another. The "weaker" reading construes externality as "not a spatial but a metaphysical notion such as ontological independence" (2020, 68).

Jacobs assumes a certain understanding of Du Châtelet's metaphysical project, which is shared by the commentators he engages (Stan, 2018; Brading, 2018, 2019). They identify the "Beings" whose diversity and union are said to constitute extension

---

we represent several existences that are different and therefore external to each other "as if in one" [*tanquam in uno*].

[15] Ruth Hagengruber cites Du Châtelet's commentary on Mandeville's *Fable of the Bees* as evidence that she "authoritatively disapproves" Locke's rejection of innate ideas (2012, 8). I agree with Karen Detlefsen that Du Châtelet's "departure from Locke on this point" does not "thrust [her] directly into Descartes's camp on the issue of nativism" (2019, 117, n. 39), and more generally that Du Châtelet's nativism takes the form of "innate rules of reasoning that can … be used to develop a metaphysics", not that of "a rich store of ideas that inform us of metaphysical truths about the essence of the external, created world" (2019, 121).

in Chapter V with the "fundamental building blocks of the world", the "simples" that are non-extended (Jacobs, 2020, 67) and, at least for Jacobs and Stan, physical (*ibid.*; Stan, 2018). On this understanding, both interpretations of "external" have metaphysical import; the spatial construal is aptly designated as "stronger" because it gives the notion an additional, epistemic, significance, as well as making it more determinate. On Brading's reading, in particular, the importance of Du Châtelet's account of extension in V.77 is that she employs it as a premise in her argument, in VII.133, that "the extended bodies of our experience arise from non-extended simples" (2019, 58; 2018, 157). As Jacobs argues, the extension that results from the diversity and union of "Beings" thus understood is not grounded purely in "mental facts" (even though minds are required, in order to "represent multiplicities [of simple beings] as extended"), but "also grounded in these multiplicities themselves" (2020, 69).[16] So this understanding of the "Beings" commits Du Châtelet to a sort of realism about extension.

I do not think, however, that this understanding of the "Beings" is forced on us. Du Châtelet introduces what Jacobs describes as the "fundamental building blocks" only in Chapter VII, in her account of "the origin of matter" or of "extension in matter" (VII.119; 1740/131, 1742/137). She argues that since "sufficient reason obliges us to state [a reason] different from what one is asking about", giving the sufficient reason for "the origin of extension"—specifically of the length, width and breadth belonging to bodies—requires "com[ing] in the end to something that is without extension, that has no particles", which is then a "simple Being". Du Châtelet concludes that "compounds, extended Beings, exist because there are simple Beings" (VII.120; 1740/133, 1742/139), which shows that even once she introduces the "simples", she does not use "Being" to refer exclusively to them. Still less, accordingly, should we assume that she uses "Being" to mean only the simples in the earlier Chapter V. It is more plausible that she uses "Being" with the meaning she has specified in Chapter III: "that which can exist, and whose determinations do not imply any contradiction, whether this Being exists, or whether it is only possible" (III.35; 1740/156, 1742/61). In this chapter, Du Châtelet speaks of physical objects such as rocks (III.36), clocks, and the human body (III.44) as "Beings".

Abandoning the restrictive understanding of "Beings" (as simples) allows us to consider Du Châtelet's account of extension outside the context of her deepest metaphysics, and even as detached from her metaphysical project in general. We are then able to understand the spatial construal of externality not as an enrichment of the metaphysical construal (that is, the construal as ontological independence), but as merely epistemic.

However, this way of understanding Du Châtelet's account of extension faces the immediate problem that it seems to make her account of space circular. If in representing beings as external to one another we represent them as spatially removed,

[16] While Brading is explicitly neutral regarding what kind of being (physical or mental) Du Châtelet ascribes to the simples, she does take it as a merit of Du Châtelet's account (2018, 160) that it avoids the need for "resources"—namely the "qualities, properties, and nature of mind"—that "lie outside the investigative reach" of the "study of matter in motion" (2018, 156).

we would seem to already be representing extension at the first step of the account of space. So our first problem, which I address in Sect. 5.3.1, is to explain how the idea of extension can be in any way prior to that of space. In the following Sect. 5.3.2. and Sect. 5.3.3, I explain in detail how the threat of circularity arises, and can be addressed.

By clarifying the relations of priority among the ideas Du Châtelet discusses in Chapter V, I hope to bring into focus how further content is joined to the idea of extension to form the idea of space. We can thus see how, on Du Châtelet's account, we form the idea of space (as continuous, homogeneous, isotropic, and unbounded) by means of our representation of particular instances of spatial relationships. My broader aim is to vindicate an epistemic reading of Du Châtelet's term "external" by showing this to be her purpose.

5.3.1 Metaphysical and Epistemic Theorizing About Space

A first step in addressing the apparent circularity is to distinguish epistemic issues regarding spatial representation from the metaphysical issues regarding relations which appear to drive, in particular, Leibniz's argument. We saw that Leibniz concludes his paragraph on the idea of space with the claim that places and spaces "consist only in the truth of relations" (V.47); this comes shortly after the summary claim that space is "an ideal thing, containing a certain order, wherein the mind conceives the application of relations" (V.47). The alternate descriptions of space in his letters give some indication what these relations are: space is an "order of coexistences" (Third Letter, §4), an "order of situations" (Fifth Letter, §104), or "that order which renders bodies capable of being situated, and by which they have a situation among themselves when they exist together" (Third Letter, §41). The most important point for our purposes is that Leibniz takes the metaphysics of space to depend at least in part on that of relations.

§47 of Leibniz's Fifth Letter, on the idea of space, also contains a passage whose importance for his view of relations is widely recognized. Here Leibniz distinguishes three ways of conceiving the "ratio or proportion between two lines L and M": as a ratio of L to M; as a ratio of M to L; "and lastly, as something abstracted from both, that is, as the ratio between L and M, without considering which is the antecedent, or which the consequent; which the subject, and which the object" (Fifth Letter, §47). Leibniz proceeds to argue that, considered in either of the first two ways, "that accident which philosophers call relation" has a subject (namely L, in the first case, and M, in the second). But considered in the third way, this relation cannot be an accident; for, Leibniz argues, there is no subject to which it can be attributed. It "cannot be said that both of them, L and M together, are the subject of such an accident; for if so, we should have an accident in two subjects, with one leg in one, and the other in the other; which is contrary to the notion of accidents". He concludes that on the third way of considering it, the relation is "out of" the [*hors des*] subjects, and "being neither a substance, nor an accident," must be "a mere ideal thing, the consideration of which is nonetheless useful" (Fifth Letter, §47).

Here Leibniz does not further explain how to consider accidents which are "in" the subjects (L and M). But his remarks elsewhere about relations and "purely extrinsic denominations" have often been read as requiring some sort of reduction of relational properties to non-relational ones (and even, as by Bertrand Russell, as denying there are any relational properties). An oft-cited such passage, which bears on our example of the ratio between lines, claims that quantity and position "are nothing but mere results, which in themselves constitute no intrinsic determination, and thus, that they are only relations which require a foundation taken from the category of quality or from an accidental intrinsic determination."[17]

Massimo Mugnai provides an illuminating contextualization of Leibniz's view within the "scholastic and late-scholastic tradition", which likewise distinguishes relational accidents (in our example, the ratio as belonging to L or M) from polyadic properties, and "subsumes" the former "under the Aristotelian category of *ad aliud* (*towards another thing*)". To avoid the supposition of inherence in multiple subjects, thinkers within this tradition took relational accidents to be "grounded on a fundamental property of a given subject", and to "allude" or "refer" to distinct properties belonging to the other relata (Mugnai, 2012, 188). (Realist thinkers still distinguished accidental forms incorporating such reference from these "fundamental" properties, while nominalist ones identified them.) Mugnai observes that Leibniz goes further (than this tradition) in a "nominalist-conceptualist" direction when he assimilates all relational accidents to "extrinsic denominations", thus denying them any inherence in a subject. and construing "*all relations* as a product of the mind" (2012, 192).

The relevant point is that whatever problems this directedness "toward another thing" presents (or may have been taken by Leibniz to present) for metaphysics, it is not as problematic for theorizing about spatial representation. For Du Châtelet and Wolff, as I understand them, what requires explanation is specifically our capacity to represent space as continuous, homogeneous, isotropic, and unbounded. Wolff's description of the process by which we come to represent space—by "attending to the coexistence of simultaneous [things] A, B, C, and D, etc.", and distinguishing the "manner" in which each thing coexists with each of the others (*Ont.* §588)—is highly abstract, to be sure. But Wolff does identify these "manners" of coexistence with distances, and further claims that the truth of his claim "is shown by experience [*patet per experientiam*], if we attend to any objects near us" (*ibid.*). It is plausible to understand the experience on which our capacity for spatial representation is based as perception of particular objects as standing in determinate spatial relations. Visual experience, in particular, can directly exhibit determinate distances and orientations, namely when the objects thus related are presented simultaneously within one's field of vision. Such instances of spatial relationships can be taken as basic for the purpose of explaining the representation of extension and of space, even if the directedness "toward another thing" that they exhibit raises puzzles for metaphysics (Leibniz, Wolff, and Du Châtelet would presumably welcome the implication that

[17] "On the Principle of Indiscernibles," quoted in Arthur 2021, 350.

our experience of the world as spatial is a poor guide for metaphysics.). This also gives the capacity for spatial representation a basis in experience, making it possible to avoid the supposition of innate ideas.

5.3.2 *Kant's Apriority Argument*

Wolff's account, as I have very briefly sketched it, remains threatened by circularity. I will show how this threat arises in order to show that Du Châtelet's account may be able to escape it, and thereby bring out the difference in their approaches.

The charge of circularity is leveled by Kant, who appears to target Wolff's account in his argument for the apriority of space.[18] Kant's use of "outside" [*ausser*] in this passage is ambiguous in the same way as Du Châtelet's use of "external", and it will be instructive to consider how Kant's interpreters have resolved the ambiguity. Kant claims that the representation of space is presupposed by (that it "must already lie at the ground" of) representation of things[19] as "outside one another, thus not merely as different but in different places" (A23/B38). Despite this explicit specification of the meaning of "outside" in spatial terms, commentators have argued[20] that it must have the strictly ontological meaning of numerical distinctness, on pain of making Kant's claim a tautology. For if "outside" is taken in its spatial sense, Kant's claim appears to be of the form "the representation of x-ness is presupposed by the representation of objects as x". Since this appears to hold equally of any number of empirical concepts, it could not support Kant's assertion of a special status (apriority) for the representation of space.

Daniel Warren has argued, however, that this representation is special precisely in that the representation of particular objects as spatially related presupposes the representation of a comprehensive spatial framework (so that the latter could not have been acquired by abstraction from the former). Warren makes this point by contrasting concepts of qualities such as brightness with the concept of space. As Warren observes, we can represent the relative brightness of a particular object by conceiving of it as occupying a "place" in a relational structure, which we may call "brightness-space". Employing such a structure to "keep track" of relations of brightness requires more from us than merely representing particular instances of comparative brightness ("brighter-than" relations):

> [I]n thinking of objects as located in brightness-space, we represent them not simply as bearing "brighter than" relations to one another, but as each bearing another relation to

---

[18] This is the first numbered argument in the section of the Transcendental Aesthetic that is titled "Metaphysical Exposition of Space" in the second edition (and simply titled "First Section" in the first edition) of the *Critique of Pure Reason*.

[19] Kant in fact speaks of what is required to represent "sensations" as outside one another (A23/B38). It is standardly assumed that he means to discuss the representation of objects to which the sensations are related. See, *e.g.*, Allison 2004, 100–1, and, for discussion, Falkenstein 1995, 160.

[20] In arguing against this interpretation, Daniel Warren attributes it to Henry Allison, D. P. Dryer, P. F. Strawson, and R. P. Horstmann (1998, 198 *n*.25).

something further: this other relation is that of occupying or being in; this something further is a place in, or a region of, a 'brightness-space.' (1998, 199-200)

Warren's point is that these additional representational elements (relations and places) are not required in the first place to represent instances of comparative brightness. He argues, further, that just these elements are required in order to represent instances of spatial relationships. There is thus

an important difference between the kinds of account we could give of how we form the representation of the brightness-[space] and the kinds of account that would be possible in the case of space (or of time). For, given that the representation of "brighter than" relations does not presuppose the representation of brightness-space, we can form the latter representation from the former. By contrast, in the case of spatial (or temporal) relations, we cannot proceed in an analogous fashion, forming the representation of space (or of time) from a concept of spatial (or temporal) relations which has been independently acquired, and more specifically, which has been independently acquired from experience. (1998, 204)

The comparison between the representation of relations such as comparative brightness and representation of spatial relations is thus intended to show that empirically perceived instances of spatial relationships cannot, after all, serve as the basis from which the representation of space is formed.

It is not clear that this argument has force against Du Châtelet's view. The argument mainly targets the explanation of how we could first come to form the representation of space through experience, and Du Châtelet (in contrast to Wolff) is not committed to finding an empirical origin for it. Her aim, as I understand it, is to clarify the relationship between the idea of space and the idea of extension, and more specifically to make clear how the content of the idea of space goes beyond that of the idea of extension. She can show that the idea of extension is more basic than that of space by abstracting out the idea of extension as an element (or necessary condition) of experience that has full-blown spatial content, without assuming a prior stage of experience in which extension alone is represented. However, Du Châtelet would still be committed to regarding the representation of extension—in particular, of things as "external to" one another—as "independently" available, to use Warren's terminology. So if the argument succeeds in establishing that any representation (empirical or otherwise) of particular spatial relationships presupposes a comprehensive framework of places, it does have force against her position.

The point of bringing Kant's argument into consideration is not to resolve this dispute but, first, to bring into relief the difference between Du Châtelet's approach and Wolff's. Secondly, consideration of Kant's argument shows that even on a purely spatial reading of terms such as "*hors de*" and "*ausser*" (such that they have no metaphysical significance), the assertion of a necessary linkage between the representation of extension (specifically, of objects as "outside" one another) and the representation of space can be more than a mere tautology. Kant's dispute with Wolff (and arguably also Du Châtelet) illustrates what is at stake when the representation of extension is asserted to be prior to the representation of space, or vice versa.

5.3.3 The Metaphysical Significance of the Notion of Externality

The point of Sect. 5.3.1 and Sect. 5.3.2 was to show how the representation of extension—that is, of things as "external to" one another (and unified)—can serve as a basis for the representation of space, even if "external" is a purely epistemic notion (that of spatial removal). In fact, however, I do not think it has exclusively spatial meaning for Du Châtelet. That it also has metaphysical significance is apparent from the way Du Châtelet argues, just after defining "extension", that the notion is misapplied by "Philosophers" who take the soul to be extended because it has "several different faculties that nonetheless constitute a single subject" (V.77; 1740/97–8, 1742/102). Her argument is that "since the attributes and modes of a Being are inseparable from it, is incoherent to consider them as external to each other" (closely paraphrased by Jacobs, 2020, 68). I take Du Châtelet's use of the terms "attribute" and "mode" here as evidence that we are to understand her use of "Being" in relation to her account in Chapter III, for that is where she introduces them, making clear that attributes and modes are determinations of "Beings" rather than "Beings" themselves. Since attributes and modes are thus characterized by their ontological dependence, we should take her to argue that attributes and modes of a common subject cannot be considered external to each other because their dependence on this subject keeps them from being ontologically distinct. Externality must thus involve ontological distinctness, but (*pace* Jacobs, Stan, and Brading) it does not have to involve the existence of "simple beings".

Before moving on, we should consider Jacobs' criticism of the spatial reading of "external". His strongest objection is that "the example Du Châtelet gives to illustrate her notion of externality fits the weak reading" (2020, 68).[21] The example is of "imagining a structure that we have never seen before".

> [W]e represent it as external to ourselves, even though we know well that the idea we have of it exists within us, and that there is perhaps nothing of this structure existing external to our idea. But we still think of it as being external to ourselves because we know that it is different from us... (V.77; 1740/97, 1742/101)

Jacobs observes that it makes more sense to understand "different from us" as ontolologically distinct than as differently located (2020, 68), so it is more plausible to understand "external" as nonspatial (whether we take the "because" clause to articulate its meaning, or to explain why we employ this mode of representation). Jacobs further observes that in the argument just discussed, it would be odd to think of attributes and modes as spatially distant from their bearer. It could be replied that the "incoherence" to which Du Châtelet points would hold on either

[21] Jacobs argues, in addition, that Wolff's influence on Du Châtelet supports the ontological reading, since Jacobs takes Wolff to "use the term 'external' in a non-spatial sense" (2020, 68). Although Jacobs does not note it, Wolff also uses this example ("*Si nobis imaginamur aliquod aedificium magnifice extructum, satis nobis conscii sumus idem esse a nobis diversum: aedificium vero istud extra nos repraesentamus ....*"; *Ont.* §544).

reading, but defenders of the spatial reading would be challenged to explain how such “Philosophers” could be taken to make the mistake in the first place.

I think we can grant these points, since (as I have just argued) “external” must have metaphysical import even on a spatial reading. I still maintain, however, that Du Châtelet is more concerned to explain how these ideas—in particular, that of extension—can be employed in science in general than to determine the ontological status of space (or of extension).

5.4 The Thinness of the Idea of Extension

We saw that on Du Châtelet’s view, we form the idea of extension by considering distinct beings as “united together” (V.77). She describes extension not only as an “abstract being” (V.79, quoted above) but as an “imaginary Being” (V.79; 1740/99, 1742/104) and as an “ideal Being” (V.79; 1740/100, 1742/104; V.80). This can easily lead us to believe that her main point is to assert the mind-dependence of extension. We could take her, moreover, to trace the mind-dependence of space to the moment (within this analysis) of unification. There are antecedents for such a view in the history of philosophy. Indeed, prominent Leibniz scholars hold that his view of relations was influenced by a “nominalist-conceptualist” tradition on which “pluralities exist only as the individuals they are aggregated from, so that the unity of an aggregate is something added in the act of perception” (Arthur, 2021, 13, citing Mugnai, 2012, 204). Such a view would also have close affinities with Kant’s view of objects (both of ordinary perception and of scientific knowledge) as constituted by the understanding’s syntheses, as when Kant defines an object as that “in the concept of which a manifold of given intuition is united” by a synthesis of the understanding (B137).[22]

But I think we should resist this tendency. It is implausible that Du Châtelet aims to specify the ontological status of extension, because the content that is excluded from the idea, on her account, is just what would be needed to determine its status. We can see this by, first, comparing the idea of extension with that of space (which does include this content), and second, considering what is needed to determine ontological status.

5.4.1 Comparing the Idea of Extension with the Idea of Space

In her final paragraph on extension, Du Châtelet emphasizes the thinness with which its parts (the diverse and unified Beings) are conceived:

[22] To be sure, Kant does not argue in this way for the mind-dependence of space, although the unity of the “single all-encompassing space” (A25/B39) does play a role in his argument.

> [W]e see that the parts of extension, when considered abstractly and without taking into account any limits or shapes, do not have to have any internal differences; they must be similar, and differ only in number. For since, in order to form the idea of extension, we consider only the plurality of things and their union …, and we exclude every other determination (all the parts being the same with respect to plurality and unity), we can substitute one in the place of another without destroying … the plurality and the unity (which are the only determinations to which we pay attention), and so any two parts of extension can differ from each other only in being two not one. So all of extension must be conceived of as … having no internal determination that distinguishes one part from another…. (V.78; 1740/98-99, 1742/103)

The paragraph concludes with the claim that "this notion of extension is still that of a geometrical body" (V.78; 1740/99, 1742/103–4). The following paragraph explains how the idea of space arises, "when we consider the possibility that many different things may exist together in this abstract being we call *extension*". On this account, the idea of space is "the idea of extension joined with" the possibility we are considering, namely that "of restoring to the coexistent and unified Beings, from which the idea was formed, the determinations that we had already stripped from them by abstraction" (V.79; 1740/100, 1742/105). Although Du Châtelet speaks of the possibility "of restoring" [*de rendre*], what goes into the idea of space should be understood not as our capacity for a process or act, but as the content we are entertaining. Considering the beings as at least possibly bearing such determinations first makes it possible to "attend solely to their manner of existing externally to one another", which then makes it possible to "represent to ourselves that this coexistence of several Beings produces a certain order or resemblance in their manner of existing" (*ibid.*).

Du Châtelet does not spell out exactly how we come to represent this "manner of existing externally" or the "order of resemblance" to which it is subject. But she appears to have in mind the account that Leibniz gives in his Fifth Letter to Clarke and Wolff revisits in *Ontologia*. On this account, we form the notion of space by first representing an order in which things coexist which are different from one another.[23] We then come to represent places in this order, as Leibniz explains.[24]

---

[23] Leibniz writes that "men consider that many things exist at once and they observe in them a certain order of coexistence" (V.47). Wolff claims we attend "to the coexistence of simultaneous [things] A, B, C, and D, etc." and, for each thing (A, B, C, …) distinguish the manner in which it coexists with each other thing (B, C, D, …) from that in which it coexists with the remaining other things (C, D, …); "insofar as they are placed next to each other in a certain order, … then we have the notion of space" (*Ont.* §588).

[24] In contrast, Wolff says that "we have the representation of space" once we attend to the distinct manner in which each of the simultaneous things coexists with each of the others ("insofar as they are placed next to each other in a certain order, such that the distance between A and C is distinct from that between A and D, etc."), without mentioning place (*Ont.* §588). But in representing these distinct "manners of coexistence", we are representing places, according to his definition later in the chapter. "While simultaneous [things] coexist in a certain order, so that the manner in which A coexists with B, C, D, etc. can be distinguished from the way in which B coexists with A, C, D, etc. and so on; it is clear that a certain manner of coexistence belongs to each of the simultaneous [things], which is different from the manner of coexistence of each of the others. What is determined in this way is the manner in which A coexists simultaneously with B, C, D, etc., which is what we call *place*" (*Ont.* §602).

> When it happens that one of these coexistent things changes its relation to a multitude of others … and that another thing, newly come, acquires the same relation to the others that the former had, we then say it is come into the place of the other … to give a kind of definition, *place* is that which we say is the same to A and B when the relation of the coexistence of B with C, E, F, G, etc., agrees perfectly with the relation of the coexistence which A had with the same C, E, F, G, etc. … (V.47)

Space, finally, is "that which comprehends all those places" (*ibid.*).

So on this account, to represent space we must represent the places. To represent the places, we must be able to reidentify "relations of coexistence", when one thing comes to have the same ("perfectly agreeing") relation that another had previously. Now if we understand this as an account of how we form the notion of space, and presumably also of place, it appears that we first become able to represent these relations (of coexistence) and their sameness ("perfect agreement") only by representing the things.[25] In that case, we would have to be able to reidentify the things in order to reidentify the relations. That explains why we must be able to represent the "existing beings" (A, B, C, D) as not only different from one another, but as having individuating characteristics. To this end, we must restore to the beings those determinations that we "stripped" in order to regard them as "differing only in number".[26]

Incidentally, these "other qualities which we have mentally separated" appear to be the "things" [*choses*] that Du Châtelet says are "received and contained in" the "ideal Being" that we first conceive as extension, "as a vessel receives liquid that is poured into it" (V.79).[27] This understanding of it as a receptive medium corresponds, more closely than the description of space as a fluid in which bodies swim (V.72), to how space is conceived in the Leibniz-Newton dispute. For Leibniz frequently speaks of extension as the "diffusion" of "something prior".[28] He sometimes speaks of that which diffuses in qualitative terms, for instance as "the continuity in which a part is

[25] The restriction to representing the relations only in this way seems to be the force of Leibniz's qualifications that we say one thing comes into the place of another when the latter "changes its relation to a multitude of others, *which do not change their relation among themselves*" and the former "acquires the same relation to the others, that the former had"; and that we may say that "those which have such a relation to those *fixed existents*", which "*have undergone no change*", "have now the same place which those others had" (V.47, emphasis added).

[26] "[S]ince, in order to form the idea of extension, we consider only the plurality of things and their union …, and we exclude every other determination (all the parts being the same with respect to plurality and unity), we can substitute one in the place of another without destroying … the plurality and the unity (which are the only determinations to which we pay attention), and so any two parts of extension can differ from each other only in being two not one. So all of extension must be conceived of as uniform, homogeneous, and having no internal determination that distinguishes one part from another…." (V.78; 1740/98–99, 1742/103).

[27] Our mind "conceives of the other qualities that we have mentally separated and that are no longer a part of our idea of this Being. Therefore, it seems as though we import all of these things into this ideal Being, [which] receives and containing them, as a vessel receives liquid that is poured into it" (V.79; 1740/100, 1742/104).

[28] Extension "is merely the repetition or diffusion of something prior" (Postscript of a Letter to Basnage de Beauval, AG 147). Similarly: "the notion of extension is incomplete in itself, but is relative to something which is extended, something whose diffusion or continuous repetition

similar to the whole, as, for example, we conceive of whiteness as diffused in milk, the same direction as everywhere diffused in a straight line, and equal curvedness as diffused in the circumference of a circle" (Letter to de Volder, AG 183).[29] And in *Specimen Dynamicum* (which was known to eighteenth-century readers), Leibniz speaks of that which diffuses in terms of the powers (of acting and reacting) that characterize substance (AG 118), which also serve to distinguish individual substances from one another. Du Châtelet could thus take herself to agree with Leibniz (although she may not interpret him correctly) that insofar as space is not merely ideal, it exists as the medium in which individuating qualities can be thought to spread or diffuse.

Of course, this does not yet explain how Du Châtelet can attribute the view of space as a fluid (in which bodies swim) to the opposed Newtonian position. However, she begins the following chapter, "Of Time", with two analogies between time and space; the second[30] is that "ordinarily our image of time, *as of space*, is produced by confused ideas" (VI.95; 1740/113, 1742/118–19, emphasis added). In the case of time, "one imagines it as a [self-subsistent] being composed of continuous, successive parts," which "flows uniformly" and continuously for "all eternity" (*ibid.*). (Du Châtelet's use of "ordinarily" and the impersonal "one" suggest that all of us—not only the proponents of substantivalism—do in fact imagine time this way. That claim would comport with her view that we have the "habit of representing our ideas with perceivable images" (VII.135; 1740/150, 1742/158), and with the general tenor of Wolffian criticisms of Newtonians, namely that they allow imaginary notions at the basis of their metaphysics.[31]) With respect to time, however, what is really continuous and uniform is just the succession of beings that exist in time,[32] and it is presumably through confusion that we attribute continuous and uniform "flow" to the "parts" of the order in which they succeed. Filling out the analogy, we may suppose that when we—or at any rate the substantivalists—confusedly imagine space as a thing, we attribute diffusion to the parts of the order in which beings coexist, as if the order itself were spreading out. In reality, what diffuses are the qualities by which individual beings, and thus places in the ordering, are distinguished.

---

extension indicates" (*Specimen Dynamicum*, AG 130; *cf.* AG 118); extension "represents only a certain … diffusion or repetition of a certain nature" (Letter to de Volder, 30 June 1704, AG 178).

[29] Similarly: "since extension is a continuous and simultaneous repetition, it follows that whenever the same nature is diffused through many things at the same time, as, for example, malleability or specific gravity or yellowness is in gold, whiteness is in milk, and resistance or impenetrability is generally in body, extension is said to have place" ("On Body and Force, Against the Cartesians", AG 251).

[30] The first analogy is that in space, "one simply considers the order of the coexistents insofar as they coexist", while in time, one considers "the order of successive things, insofar as they succeed each other" (VI.94; 1740/113, 1742/118).

[31] See Dunlop 2013.

[32] Du Châtelet speaks of "continuous succession" of beings at VI.97 and VI.101. On her account, the succession of beings is uniform in the sense that there is "no possibility of the existence of another being" between any two succeeding beings; this leads us to conceive of "an ideal being, consisting in a uniform flow, which must be similar in all its parts" (VI.98; 1740/117, 1742/122).

5.4.2 What is Required to Determine Whether Beings Can Exist

In any event, we now see that the idea of extension differs from that of space in that it excludes, while the latter includes, even "the *possibility*" of restoring determinations "stripped by abstraction" from the "coexistent and unified Beings" (V.79, emphasis added). So to conceive extension properly (in a manner that respects the difference between it and space), it is necessary to conceive the beings that constitute it as having no other determinations than their diversity and union.

I contend that considering the beings in such a way makes it impossible to determine whether, or under what conditions, they can exist (and thus whether or under what conditions they can coexist in a unified manner, *i.e.* extension can exist). It becomes impossible, in particular, to apply the criteria of possibility presented in Chapter III, which determine the application of the notion of "Being" as Du Châtelet explains it there: "that which can exist, and whose determinations do not imply any contradiction, whether this Being exists, or whether it is only possible" (III.35; 1740/56, 1742/61). This is the only understanding of "being" available at this stage of discussion, at least on my reading.

Du Châtelet tells us in Chapter III that "insofar as a Being is considered alone, one must show its intrinsic possibility by the principle of contradiction, and its external possibility, or actuality, by the principle of sufficient reason" (III.50; 1740/69, 1742/74). To show intrinsic possibility, we must show that the thing "contains nothing that is contradictory" (as, for instance, a triangle is possible, "because it is not at all contradictory that three lines may be joined at their ends and enclose a space"; III.34; 1740/55, 1742/60). But since we cannot consider the "coexistent and unified" beings as having *any* internal determinations, we cannot complete this first step. Du Châtelet claims, further, that "when one considers [a] Being as placed in the order of things, … one must show how a Being depends on its neighbor, and which causes gave actuality to the modes that were simply possible when the Being was considered as isolated and outside the order of things" (III.50; 1740/70, 1742/74). Since the actuality of modes "depends either on other antecedent modes, or on exterior Beings, or on both at the same time" (III.44; 1740/62, 1742/68), one would have to conceive the beings in question as possessing at least some particular modes, and perhaps also as participating in transuent causation. But, clearly, we conceive the "coexistent and unified" beings too thinly to fulfill these conditions. In short, extension is conceived as something not subject to ontological classification; space, in contrast, is conceived as something that we mistakenly classify as a substance.

5.4.3 *Why Du Châtelet Claims Extension Appears to Us as a Substance*

I will go on to suggest that extension is apt to illustrate the role of abstractions in science precisely because we conceive it too thinly to determine its ontological status. But the claim that we conceive extension so thinly faces the difficulty that Du Châtelet says more than once that *extension* "appears" to us as a substance, when on my reading only *space* could appear to us as having any particular ontological status. The main such passage, however, begins by glossing "extension" as "this ideal Being which we form from the plurality and unity of all these Beings" (V.80; 1740/101, 1742/105), and "these" presumably refers back to the closing sentence of the preceding section, which is about space rather than extension.[33] Furthermore, the passage concludes that "we are led to represent Space as a substance independent of the Beings that are placed in it" (*ibid.*), and the argument for this claim seems to apply to space rather than to extension. Here Du Châtelet draws on her earlier explication of substance as "*that which conserves the essential determinations and the constant attributes, while the modes in it vary and succeed one another*, that is to say, a durable and modifiable subject" (III.52; 1740/73, 1742/77). Accordingly, her argument proceeds by showing, first, that the "ideal Being" appears enduring (in the sense of retaining possession of certain "essential" determinations), and, second, that "this Being" appears modifiable. The first step is the claim that we "imagine several things existing together, and stripped of all internal determinations", which appears equally applicable to space and to extension. But the second can apply only to space: "insofar as it is possible by an act of our understanding, to restore to these Beings the determinations that we have stripped from them by abstraction, it seems to the imagination that we are importing something that had not been there before; and so this Being appears to be modifiable" (*ibid.*).

We may conjecture that Du Châtelet speaks of "extension" at the start of this passage (V.80) in order to streamline the argument of later chapters. For she reiterates that extension "must seem to us a Substance" at VII.134 (1740/149, 1742/156)[34] and at VIII.150 (citing VII.134). In these chapters, extension is spoken of as a feature of matter and bodies. It is thus conceived more determinately than on the account of Chapter V, which we saw to involve abstraction from solidity and whatever other features distinguish bodies from geometrical objects.

---

[33] "For the idea of Space arises from our attending solely to their manner of existing externally to one another, and representing to ourselves that this coexistence of several Beings produces a certain order or resemblance in their manner of existing; and so once one of these Beings is taken to be the first, another becomes the second, another the third, and so on" (V.79; 1740/100, 1742/105). On my reading, what makes this applicable to space rather than extension is that we take into account the coexisting "Beings"' "manner of existing", rather than considering them solely as "diverse and united".

[34] At VII.134, Du Châtelet refers back to III.52, which asserts that "every being that endures and is modifiable is a substance". She is clearly also relying, if only implicitly, on the argument of V.80 (discussed in the preceding paragraph of the main text), in which the "ideal Being" is first referred to as extension, and subsequently as space.

One reason for Du Châtelet to adopt this different usage of "extension" is to match that of "the Cartesians" (whose views Du Châtelet is criticizing) and of Leibniz (whose explanation of extension she endorses).[35] There may be a further reason for her to speak of "extension" rather than "space". For she goes on to make the point that if we could "see distinctly all that composes extension, this appearance of extension that falls under our senses would disappear and our Soul would perceive only simple Beings existing external to one another" (VII.134; 1740/149, 1742/156). First of all, the appearance in question is that of the length, width, and depth that belong to individual bodies (VII.120). Neither this appearance, nor the distinct conception which she now puts forward to explain it, involve consideration of the manner in which beings coexist; hence, neither is a representation of space. Secondly, the latter, distinct conception is a representation of the coexisting beings as not merely coexistent and unified, but simple. Since it requires us to conceive the beings in this particular way, rather than just considering it possible to "restore determinations" to them, it goes further beyond the (proper) conception of extension than does the conception of space. Du Châtelet acknowledges what she calls our "reluctance" to conceive "how simple beings without extension can by their assemblage make up beings with extension" (VII.135; 1740/150, 1742/158). To put her account of the "appearance" of the three-dimensional extent of bodies in terms of space would invoke the possibility of giving determinations to both the apparent and the underlying beings, and would thus make salient that the latter must be either composite or simple. Since the apparent beings cannot be conceived as simple, this would confront us directly with the puzzle of how they can be "made up" of simple beings. By speaking instead of extension, Du Châtelet is able to keep this puzzlement further in the background.

5.5 The Example of the Line

On my interpretation, Du Châtelet is mainly concerned to explain the positive epistemic role of the idea of extension, and of the manner of thinking that gives rise to it. We can begin to understand the positive role of this representational capacity by considering the example she uses to illustrate it.

Immediately after defining "extension", Du Châtelet claims that we "give extension to a line, insofar as we pay attention to several distinct parts which we see as existing externally to one another, which are united together and are for this reason a single whole" (V.77; 1740/97, 1742/102).[36] At first blush, it is unclear how this

[35] We may note here that in his letters to Clarke, Leibniz claims extension "must be the affection of some thing extended", and goes on to argue that it cannot be an attribute of empty space (IV.9). Likewise, at V.46 he insists on the distinction between "the extension of things" and "the space according to which that extension is taken": "Things keep their extension; but they do not always keep their space. Every thing has its own extension, its own duration…".

[36] Du Châtelet's claim that "we see" [*nous voyons*] the line's parts "as [*comme*] existing externally to one another" might seem to call for a spatial reading of "external", since ontological distinctness

example furthers Du Châtelet's purpose. Since Du Châtelet does not characterize extension as unlimited or all-encompassing, representing a line is tantamount to representing extension. It thus appears that if we are in position to grasp this example, we must already be operating with the representational capacity that is here being explained. So to understand Du Châtelet's strategy, we must see how this particular instance of extension is suited to illustrate the properties she attributes to extension in general.

The first way in which the line is an apt example is that Du Châtelet classifies it as a geometrical, not a physical, object. For as she argues in the later chapter "Divisibility and Subtlety of Matter", only solids exist "in nature":

> Every body has ... three dimensions, or to speak with exactness, there are nothing but solids in nature. But since our mind has the power to make abstractions, we can consider length without taking note of width nor depth, ... and geometry is founded on these abstractions of our mind. Surfaces, lines, and points thus are not matter, but one conceives them in matter by abstraction. (IX.166; 1740/180, 1742/189)

The line thus illustrates, first of all, that the "parts" whose diversity and union constitute extension need not be articulated prior to representing them as extended. Du Châtelet makes explicit that the division of geometrical objects into parts is indeterminate.

> The geometrical body is nothing but simple extension. The geometrical body has no determinate and actual parts, it contains nothing but simply possible parts, which one can increase to infinity as one wills; because the notion of extension contains nothing but parts coexisting and united, and the number of these parts is absolutely indeterminate, and does not enter into the notion of extension. (IX.169; 1740/181, 1742/190)

Du Châtelet also describes the division of geometrical objects into parts as "ideal" (IX.171; 1740/185, 1742/193), which indicates that these modal distinctions apply to divisions or separations made only in thought. We thus see that one can represent parts as "external to"—meaning at least, ontologically distinct from—one another without having to actually distinguish or separate them (even in thought), as long as such separation is possible.

A still more important respect in which the line is an apt example is that, as "nothing but simple extension", it illustrates the simplicity of the idea of extension, as compared with that of space. We have seen that the idea of space differs from the idea of extension in that it includes the "possibility of restoring" just the determinations that we "exclude" from thought when thinking of extension. In the case of the line, in particular, there is no consideration of any other determination than length. The line recurs as an example in Du Châtelet's discussion of the mind's power of "forming, by abstraction, imaginary Beings that contain only the determinations we want to examine, and excluding from these Beings all other determinations by means of which they can be conceived in another manner". Thus when we wish "to measure a

---

is presumably not something we can see. However, the conjunction "*comme*" can also be translated as "as if" or "like". So we could take her to describe the way in which we see the parts as a visual representation "as if" of (not literally of) ontological distinctness.

distance, we can represent it to ourselves as a Line with neither width nor thickness, and without any internal determination" (V.86; 1740/105, 1742/110–11). We can easily see how the representation of extension is better suited to illustrate this power than that of space. For the determinations by which beings "can be conceived in another manner" (than merely as diverse and unified) are strictly excluded from the representation of extension; but the representation of space requires at least the possible inclusion of these determinations (so as to distinguish the beings and determine when any two of them bear the same relation(s) to the other(s)).

5.6 The Usefulness of Abstractions

An immediate aim of Chapter V is to explain space as a product of abstraction which nonetheless appears to us as "self-subsistent, penetrable, immutable, infinite, etc." (V.85; 1740/105, 1742/109). But the marginal title of V.86—"Usefulness of Abstractions"—and its placement in the overall organization of Chapter V show the account of mental abstraction to have broader importance. Du Châtelet claims to have completed her account of space in V.85, which begins "Here, then, is the origin of all the properties that we attribute to Space …" and continues, "… we see that all of these alleged properties … have no reality but in the abstractions of our mind" (1740/105, 1742/109). Indeed, the term "space" does not occur in V.86; and when Du Châtelet returns to discussing space, in V.87, she relies on points made about extension, or so I argue below.

On a larger scale, the placement of Chapters V and VI within *Institutions* shows how the mental power of abstraction takes on importance at this juncture. Katherine Brading has argued that the overall organization of *Institutions* follows that of "Newtonian texts" such as Keill (1720), 's Gravesande (1720), and Musschenbroek (1744), which begin with "considerations of method" (Brading, 2019, 6) and move on to such topics as body (considered with regard to extension and solidity); motion, place and time; laws of nature; and such phenomena as descent on an inclined plane and oscillation of pendula. We indeed see this structure in Chapters VII through XVIII of *Institutions*, which cover "the bodies that are the subject matter of physics" (Chapters VII–X, quoting Brading 2019, 7); laws of motion (Chapters XI–XII); gravitation (Chapters XIII–XVI); and descent on inclined planes and oscillation (Chapters XVII–XVIII). Brading's claim that Chapters I through IV should all be considered "primarily methodological in intent" (*ibid.*) is controversial; but this clearly holds of Chapter IV, on hypotheses, so on any account, there is a question of how Chapters V and VI enable the transition from methodological discussion to physical theory.[37] I now want to show how the discussion of abstraction can be seen to continue the account of hypotheses and their use, specifically by bringing to light a positive use of "fictions".

[37] On Brading's own interpretation, space and time are treated as "prerequisites for an account of extended bodies and their motions" (2019, 7).

After she gives the examples of a line and two-dimensional extension to illustrate how we represent beings "contain[ing] only the determinations we want to examine", Du Châtelet asserts that "these fictions" are useful for discovering new truths and relations, as long as "we do not imagine that there exists anything resembling these abstractions of our mind" (V.86; 1740/106, 1742/111).

If we focus too narrowly on this proviso, we may fail to see how Chapter V goes beyond Leibniz's and Wolff's treatment of space, as well as the earlier chapters of *Institutions*. I find two previous uses of the term "fiction" in *Institutions*, both (in Chapter IV) to describe uses of hypotheses that do not meet the standards of scientific rigor.

> When the necessary consequences do not follow from [a supposed hypothesis], and a new hypothesis must be created to use the first to explain the phenomenon, this [first] hypothesis is only a *fiction* unworthy of a philosopher. (IV.69; 1740/88, 1742/92, emphasis added)
>
> By distinguishing between the good and the bad use of hypotheses, both extremes are avoided, and without giving oneself up to *fictions*, a method very necessary to the art of invention is not denied to the sciences ... (IV.71; 1740/88, 1742/92-3, emphasis added)

In V.86, the proviso that we should not suppose "that there exists anything resembling" the beings we conceive by abstraction is followed by similar conditions on when "imaginary" notions can be substituted for "real" ones:

> ... we must resign ourselves to using these imaginary notions whenever we can substitute them for real ones without prejudicing truth, just as we use Ptolemy's system to resolve many Problems of Astronomy where the solution becomes much more difficult when using Copernicus's system, because we can in these instances substitute one hypothesis for the other without damaging the truth. (1740/106, 1742/112)

We could take the thrust of this passage to be the negative point that the use of "imaginary notions" is impermissible when it "damages" truth, as in metaphysics and presumably also physics (in contrast to predictive astronomy). We would then see the discussion of space as reinforcing the point made in Chapter IV, that fictions should not be confused with notions of real things in philosophy. Such a reading would make sense of why the discussion of abstraction comes immediately after the summary claim that the alleged properties of space, "as well as the Being in which we suppose them to exist, have no reality but in the abstractions of our mind" (V.85; 1740/105, 1742/109).

Yet further consideration suggests that Du Châtelet aims to do more than eliminate the confusion between space as an idea formed by abstraction and the realities with which philosophy deals. Not only is the term "space" absent from V.86, but the term "fiction" here has a more positive valence than in IV.69 and IV.71, as Aaron Wells has observed (2023, 85). The very title "Usefulness of Abstractions" puts emphasis on how fictions serve to advance knowledge, and the main text identifies two specific ways in which they are useful, subject to the proviso just noted. As the example of the Ptolemaic system illustrates, assuming a fictional hypothesis can make it easier to solve problems; and fictions lead to novel discoveries by allowing the mind to operate at the abstract level, in cases where its subject-matter would be too difficult to contemplate "in the concrete".

> Provided that we do not imagine that there exists anything resembling [them], these fictions help in finding new truths and new relations, for our mind rarely has enough force to contemplate that which is Abstract in the Concrete without being distracted by the multiplicity of things that it must represent to itself. Also for all of the Sciences, and especially Mathematics, are they not full of these sorts of fictions which are one of the greatest secrets of the art of invention … ? (1740/106, 1742/111)

Du Châtelet's claim that fictions are of use in "all the Sciences" cancels the suggestion that they are appropriate only for mathematical sciences such as predictive astronomy.

These considerations allow us to see a way in which Chapters V and VI enable the transition from methodological considerations to properly physical ones. We may conjecture that Du Châtelet follows the discussion of hypotheses in Chapter IV with this material on space and time because it affords such a good opportunity to clarify her view on fictions. Once she shows by example how it can be useful to "form, by abstraction, imaginary Beings that contain only the determinations we want to examine" (V.86), she is in better position to explain how we can take imaginary, purely geometrical objects and properties as proxies for physical objects and properties (for discussion and examples, see Wells, 2023, §4.4.1).

5.7 Conclusion

The example that is suitable to illustrate the usefulness of fictions is that of extension (the idea of a line, or two-dimensional extension). In representing extension, we conceive the "diverse and unified beings" too thinly to grasp what it would be for them to coexist in a unified manner. So we are left with no conception of how something "resembling" extension could exist, and thus at no risk of taking anything in reality to resemble the idea. The question of whether "these exists anything resembling" arises—and invites an incorrect answer—only for space.

Acknowledgements An earlier version of this paper was presented at the 2023 Leibniz Society of North America meeting. I thank the participants for helpful discussion and Michael Futch for his generous comments. I am grateful to the editors of this volume, and especially to Jeff McDonough for extremely helpful comments and for sharing unpublished work.

References

Allison, H. (2004). *Kant's transcendental idealism*. Yale University Press.
Arthur, R. (2021). *Leibniz on time, space, & relativity*. Oxford University Press.
Brading, K. (2018). Émilie Du Châtelet and the problem of bodies. In E. Thomas. (Ed.), *Early modern women on metaphysics* (pp. 150–168). Cambridge University Press.
Brading, K. (2019). *Émilie Du Châtelet and the foundations of physical science*. Routledge.

Brading, K., & Lin, Q. (2022). Du Châtelet on absolute and relative motion. In C. Soto (Ed.), *Current debates in philosophy of science* (pp. 37–59). Springer.

Brandt, A. (2011). Wolffs Raum- und Zeittheorie zwischen Leibniz, Newton und Kant. In F. Fabbianelli, J.-F. Goubet, & O.-P. Rudolph (Eds.), *Wolffiana IV: Zwischen Grundsätzen und Gegenständen* (pp. 143–154). Olms.

Brown, G. (2004). […] et je serai tousjours la même pour vous: Personal, Political, and Philosophical Dimensions of the Leibniz-Caroline Correspondence. In P. Lodge (Ed.), *Leibniz and his correspondents* (pp. 262–292). Cambridge University Press.

Detlefsen, K. (2019). Du Châtelet and Descartes on the roles of hypothesis and metaphysics in natural philosophy. In E. O'Neill & M. P. Lascano (Eds.), *Feminist history of philosophy: The recovery and evaluation of women's philosophical thought* (pp. 97–127). Springer.

Du Châtelet's *Institutions de physique* is cited by section number and page number of the 1740 (Paris) and 1742 (Amsterdam) editions. Wolff's *Ontologia sive Philosophia Prima* is cited in the 1736 edition, reprinted in 1962 in the Olms Verlag *Christian Wolffs Gesammelte Werke* (Hildesheim). Except as noted, Leibniz's texts are cited as they appear in *Philosophical Essays*, Roger Ariew and Daniel Garber (eds.), abbreviated AG (Indianapolis: Hackett, 1989). Kant's *Critique of Pure Reason* is cited, following standard practice, according to page number of the 1781 ("A") and 1788 ("B") editions.

Dunlop, K. (2013). Mathematical method and Newtonian science in the philosophy of Christian Wolff. *Studies in History and Philosophy of Science, 44*, 457–469.

Falkenstein, L. (1995). *Kant's intuitionism.* Univ. of Toronto Press.

Hagengruber, R. (Ed.). (2012). *Emilie du Châtelet between Leibniz and Newton.* Springer.

Hagengruber, R. Emilie du Châtelet between Leibniz and Newton: The Transformation of Metaphysics. In Hagengruber (Ed.) (pp. 1–60).

Hutton, S. (2012). Between Newton and Leibniz: Emilie du Châtelet and Samuel Clarke. In Hagengruber (Ed.) (pp. 77–96).

Jacobs, C. (2020). Du Châtelet: Idealist about extension, bodies and space. *Studies in History and Philosophy of Science, 82*, 66–74.

Janiak, A. (2008). *Newton as philosopher*. Cambridge University Press.

Janiak, A. (2012). Newton and Descartes: Theology and natural philosophy. *Southern Journal of Philosophy, 50*, 414–435.

Keill, J. (1720). *An introduction to natural philosophy*. H. W.

Lin, Q. Manuscript. Du Châtelet on the Representation of Space.

McDonough, J. Manuscript. Émilie Du Châtelet Against Spatial Metaphysics.

Meli, D. B. (1999). Caroline, Leibniz, and Clarke. *Journal of the History of Ideas, 60*, 469–486.

Mugnai, M. (2012). Leibniz's ontology of relations: A last word? *Oxford Studies in Early Modern Philosophy, 6*, 171–208.

Musschenbroek, P. van. (1744). *The elements of natural philosophy.* J. Colson (transl.). J. Nourse.

Newton, I. (1999). *The Principia* (I. B. Cohen & A. Whitman, Trans. and Eds.). University of California Press.

Newton, I. (2004). *Philosophical writings.* In A. Janiak (Ed.) Cambridge: Cambridge University Press.

Sedley, D. (1982). Two conceptions of vacuum. *Phronesis, 27*, 175–193.

's Gravesande, W. (1720). *Mathematical elements of physicks.* John Keill (transl.).

Stan, M. (2018). Emilie du Châtelet's metaphysics of substance. *Journal of the History of Philosophy, 56*, 477–496.

Suisky, D. (2012). Leonhard Euler and Emilie du Châtelet. In Hagengruber (Ed.), (pp. 113–156).

Warren, D. (1998). Kant and the apriority of space. *Philosophical Review, 107*, 179–224.

Wells, A. (2023). In nature as in geometry: Du Châtelet and the Post-Newtonian debate on the physical significance of mathematical objects. In W. Lefèvre (Ed.), *Between Leibniz, Newton and Kant. Boston studies in the philosophy and history of science 341* (pp. 69–98). Springer.

Chapter 6
Du Châtelet, Wolff, and Leibniz on Divine Eternity

Jeffrey K. McDonough

6.1 Introduction

This essay explores Émilie Du Châtelet's neglected understanding of divine eternity and its relation to the views of Christian Wolff and G.W. Leibniz. It is divided into four parts. The first part sets the stage by distinguishing three views on divine eternity: atemporal, sempiternal, and permanent. The second part highlights an apparent tension in Du Châtelet's discussion of divine eternity in her *Institutions de physique* (1740), subsequently entitled *Institutions physiques* (1742) (hereafter *Institutions*).[1] It argues that that tension points towards Du Châtelet's acceptance of a permanent view of divine eternity. The third part argues that Wolff holds a position similar to Du Châtelet's, that is, that Wolff also holds a permanent view of both essences and divine eternity. The fourth part argues that although Leibniz may have held a permanent view of divine eternity in his youth, by the time of his famous correspondence with Samuel Clarke, he appears to have shifted away from Du Châtelet and Wolff towards an atemporal view. The essay concludes by suggesting that while Du Châtelet's views on divine eternity may indeed be considered Leibnizian insofar as she rejects Newton's and Clarke's sempiternalism, her overall position with respect to divine eternity is closer to the views of Wolff than to those of Leibniz himself.

[1] I have used Du Châtelet (1742) as the default version of her *Institutions*. Unless noted, I am responsible for all translations in what follows. I have benefited, however, from, and been frequently guided by, the translations of Du Châtelet (1740) made by Judith P. Zinsser and Isabelle Bour in Du Châtelet (2009) as well as by Katherine Brading, et. al. in Du Châtelet (2018). I have similarly benefited from translations of Wolff's *German Metaphysics* in Dyke (2019) as well as unpublished translations of Wolff's *Ontologia* by Bryce Gessell. I'd like to thank Clara Carus for her helpful comments on an earlier draft of this paper.

J. K. McDonough (✉)
Department of Philosophy, Harvard University, 202 Emerson Hall, Cambridge, MA 02138, USA
e-mail: jkmcdon@fas.harvard.edu

C. Carus and J. K. McDonough (eds.), *Émilie Du Châtelet in Relation to Leibniz and Wolff*, Women in the History of Philosophy and Sciences 23,
https://doi.org/10.1007/978-3-031-80541-7_6

6.2 Three Views on Divine Eternity

Discussions of divine eternity are often framed by a dichotomy between atemporal and sempiternal views. Thus, for example, in the introduction to his recent volume *Eternity: A History* in the Oxford Philosophical Concepts series, Yitzhak Melamed writes, "Over the two-and-a-half-millennia history of western philosophy we find various conceptions of eternity, yet one sharp distinction between two notions of eternity seems to run throughout this long history: eternity as timeless existence, as opposed to eternity as existence at all times" (2016, 1). Likewise, in the first section of her *Stanford Encyclopedia of Philosophy* article on "Eternity in Christian Thought" (there is no entry for eternity on its own), Natalja Deng begins by clarifying that "Broadly speaking, there have been two rival views of what God's eternality consists in. On the first, God is timeless; on the second, … God is in time and thus exists at every time" (2023). This way of framing the topic assumes a dilemma. God must either exist outside of time or God must exist at all times. God must either be *atemporal* or God must be *sempiternal.*

As Robert Pasnau has recently argued, however, the standard dichotomy between atemporal and sempiternal views of eternity appears to overlook "what is not only an intrinsically plausible conception of eternality, but what might also be regarded as its most historically prominent construal" (2011, 11). As will become clear, the third alternative Pasnau has in mind overlaps in potentially confusing ways with both horns of the now standard dichotomy between atemporal and sempiternal views. Pasnau fleshes out what I will call the *permanent* view of divine eternity primarily in mereological terms. He suggests that for many earlier thinkers, an eternal being is a being that enjoys a holochronic existence, that is, that "exists as a whole, all at once, for all of its existence, and does not partly exist at different times" (2011, 11). Pasnau's core idea is that, on the permanent conception, an eternal being does not have distinct temporal, or durational, *parts.* God does not have a temporal part that is, say, young and strong, and a distinct temporal part that is old and frail. God exists only as a whole, immutably, completely, necessarily.

We can get at the same notion of divine eternity by appealing to the traditional distinction between successive and permanent being. Aristotle had proposed that time itself is a purely successive being since there is nothing necessary, unchanging, or complete about time (1983 [*Physics* IV.10, 218a5-8] 41; cf. 1983 [*Physics* IV.10, 220a19] 45–46). Echoing Aristotle, Nicholas Oresme, in the fourteenth century, similarly maintains that time is "not permanent but [is] in a continuous flux and transition" (Pasnau, 2011, 379). Some three centuries later, Leibniz affirms the same point, writing to Clarke, "Whatever exists of time … being successive, perishes continually … Nothing of time does ever exist but instants, and an instant is not even itself a part of time" (2000, [5:49] 48). Ordinary creatures, partially contrasted with time, were held to be partially successive, partially permanent beings. They are successive beings—like time—insofar as their accidents are in flux and transition. A flower, for example, is a successive being insofar as it blossoms in the morning and wilts in the afternoon. They are nonetheless—unlike time—permanent beings insofar as they have necessary, unchanging, complete essences. The flower may blossom and wilt, but its essence endures through time without contingency, change, or incompleteness. Since God alone was held to be identical to his essence, it was universally allowed that God is the only purely permanent being, and hence, on this way of thinking about eternity, that God alone is purely, or entirely, eternal.

As Pasnau notes, God's "holochronicity itself might be considered a kind of timelessness," that is, God's being holochronic is itself sufficient for thinking that there is a sense in which God is *atemporal*. Pasnau argues that "Since such a being is entirely changeless, without even a distinct past or future, it fails to be located in time in the usual way" (2011, 14). If a temporal being is a being that has a distinct future ahead of it, a distinct past behind it, and is in continuous transition from its past to its future, then God cannot be held to be a temporal being. In an important, intuitive sense, a holochronic being must be atemporal. The same point can be made in terms of permanence, that is, we can likewise say that God's being permanent is itself sufficient for thinking that there is a sense in which God is atemporal. Earlier philosophers commonly took time to be essentially related to change. Partially permanent, partially successive beings like flowers and farmers endure through time in virtue of their permanent essences. In this respect they differ from time itself. Nonetheless, in virtue of their contingent, mutable, incomplete accidents, they change and participate per se in time. They perdure through time. To borrow language used by Du Châtelet, creatures are in time in virtue of their participating in the order of successive being. God, in contrast, as an entirely necessary, entirely immutable, entirely complete being, cannot similarly participate per se in the order of successive being at all. God, in virtue of being a purely, or entirely, permanent being cannot participate in time in the way that creatures do, that is, God cannot participate in time per se. We arrive at the same conclusion as Pasnau via a different route: God's existence, in an important sense, must be atemporal.

God's being atemporal in the sense just noted does not, however, entail that he must be conceived of as being entirely "apart" or "outside" of time in the way that contemporary philosophers are often inclined to imagine. There are again different ways of getting at this point. Borrowing contemporary terminology, we might note that even if God is held not to participate in time in virtue of being a successive being, he might nonetheless be held to endure—without intrinsically changing—through time. On this route, God might be imagined to be analogous to a point passing *through* time, where time is conceived of as a two-dimensional line. Just as the point has no spatial parts, so God has no temporal parts, but just as the point may pass through the length of the line, the line itself having spatial parts, so God might pass through the length of time, time itself having temporal parts. We might note that temporal existence in this sense is consistent with a being's existing only for a finite period of time. In theory, a being could endure for only ten, twenty, or thirty years. The assumption that God's existence is necessary, however, entails that if God endures through time, then God endures through all time. That is, given the assumption that God's existence is necessary, it follows that there is a sense in which God may not only be said to be in time, but also to be at all times. That is, there is a relevant sense in which God might be reckoned to exist not only atemporally but also sempiternally.

We can get at the same idea via the common thought that God's relation to time is analogous to his relation to space. It was once widely held that God is spatially omnipresent in the universe by wholly existing at every place in the universe in much the same way as Descartes famously suggests that souls are omnipresent in their bodies by wholly existing in every place within their bodies (Pasnau, 2011, 18–19). This thought offers another route into appreciating how God could be held to exist, in a sense, atemporally, and yet also, in a sense, sempiternally. For just as God could be thought to not exist spatially in the sense of not being extended through space, *parte ex parte*, in the way that bodies exist in space, so God could be thought to not exist temporally in the sense of not being extended through time, *parte ex parte*. If we restrict the predicate "spatial existence" to the kind of existence in space enjoyed by ordinary, extended beings, we may say that God is not in space. Analogously, if we restrict the predicate "temporal existence" to the kind of existence enjoyed by ordinary, successive beings, we may say that God does not exist in time, that is, that God exists atemporally. Denying that God exists in space parte ex parte, however, is perfectly consistent with holding that there is nonetheless a perfectly good sense in which God does exist in space, indeed, with a sense in which God can be held to be omnipresent. In an analogous way then, one could deny that God exists in time parte ex parte and still hold that there is a sense in which God does exist in time, indeed at every time. That is, one could similarly and consistently hold that God exists sempiternally. Thinking through commonly held views concerning God's relationship to space thus provides another route to the conclusion that one might coherently maintain that, in one sense, God exists atemporally and yet, in another sense, God exists sempiternally.

It is often difficult to know precisely what view of divine eternity an historical figure is defending. Discussions of divine eternity are often metaphorical and gesturing. If the current discussion is correct, however, the situation is even more dire than it might have seemed. For even a seemingly clear indication of an atemporal, or of a sempiternal view might be ambiguous. For a seemingly clear indication of an atemporal view might, in fact, be an attempt to emphasize one aspect of the permanent view. The suggestion that God is not in time might indicate either that God exists outside of time altogether or that God does not exist in time in the way that creatures exist in time. Likewise, a seemingly clear indication of a sempiternal view might, in fact, be an attempt to emphasize a different aspect of the permanent view. The suggestion that God exists at all times might indicate that God perdures through all time or that God endures through all time. The situation is regrettable, but not, I think, hopeless. For even acknowledging the potential ambiguities, I think we can nonetheless identify two signs—one more decisive, one less decisive—that a philosopher is committed to a permanent view of divine eternity.

The first indication is precisely a willingness to allow *both* that God, in a sense, might be said to not exist in time and that, in another sense, God might be said to exist at all times. The atemporalist should, of course, accept the former but must deny the latter. She should insist that God does not exist in time but must oppose the suggestion that God exists at all times. Conversely, the sempiternalist should accept the latter but must deny the former. She should insist that God exists at all times but must oppose the suggestion that God does not exist in time at all. Only the proponent of the permanent view is committed to both suggestions, properly understood. For, to repeat, on her view, we may say that God does not exist in time in the sense that he is in no way a successive being and so does not participate per se in the order of time. And, yet on her view, we may also say that God exists at all times in the sense that he endures through all times without beginning or end. A willingness to say both that, in a sense, God does not exist in time, and that, in another sense, God must exist at all times, is thus, I think, a decisive indication that one holds a permanent view of divine eternity as opposed to an atemporal or a sempiternal view of divine eternity. For ease of reference in what follows, I'll call this decisive indication, the *permissiveness indication.*

The second indication is a commitment to a certain order of explanation or grounding between God's nature and time. The atemporalist and the sempiternalist can reasonably be read as holding that God's eternity is essentially grounded in his relationship to time. The atemporalist holds that God is eternal *because* he exists outside of time. On the atemporalist view, that's what it is for God to be eternal. Similarly, the sempiternalist can reasonably be read as holding that God is eternal *because* he exists at all times. On the sempiternalist view, that's what it is for God to be eternal. The proponent of the permanent view, however, is more likely to see God's relation to time as being a consequence of his eternity, rather than its foundation or ground. Just as earlier thinkers typically saw time as being grounded in successive being itself, proponents of the permanent view of divine eternity typically saw the senses in which God may be said to be atemporal and sempiternal as being grounded in God's permanent being. On the permanent view, God is not outside of time (in the sense in which God is outside of time) because of the relation he bears to time, but rather because his existence is necessary, immutable, and complete. Likewise, on the permanent view, God is not sempiternal (in the sense in which God is sempiternal) because of the relation he bears to time, but rather because his existence is permanent. Although there is admittedly more room to maneuver on this point, nonetheless I think that an emphasis on God's necessary, immutable, complete nature explaining his relation to time, rather than God's eternity being explained by his relation to time, is a second, strong, albeit not decisive, indication of a permanent view of divine eternity. For ease of reference, I'll call this second indication, the *grounding indication.*

6.3 Du Châtelet on Divine Eternity

Du Châtelet's discussion of divine eternity in Chapter 6 of the *Institutions* comes on the heels of her discussion of time. I've argued elsewhere that Du Châtelet's thinking about time is effectively structured by three theses (McDonough, forthcoming).[2] The first thesis is that we have a foundational idea of successive being. The notion of successive being is I think most easily approached by analogy to Descartes's understanding of spatial extension. Descartes maintains that bodies are primitively extended in length, breadth and depth. For him, the spatial extension of bodies serves as a foundation for related spatial notions such as place, space, and distance. Du Châtelet similarly maintains that creatures have a primitive temporal dimension. For her, the temporal dimension of creatures serves as a foundation for related temporal notions such as *before*, *after*, *duration*, and *time*. Today, it is often assumed that time must either be analogous to substantival space or to relational space. That is to say, today, it is often assumed that time must either be something independent of things which are in time or be reduced to relations holding between entities (for example, to causal relations). Importantly, Du Châtelet embraces neither of those views. For her, time—more precisely temporality—is an aspect of creatures themselves. Just as Descartes takes bodies to be spatially extended per se, Du Châtelet takes creatures to be temporally successive per se.

The second thesis is that time just is "the order of successive things, insofar as they succeed each other, abstracting from any other internal quality than simple succession" (Du Châtelet, 1742, 94). The analogy to Descartes's views on extension is helpful here as well. Bodies for Descartes have a kind of spatial ordering simply in virtue of their being extended per se. Their spatial ordering is not dependent upon their being related to an independent, absolute space, nor even to their standing in further, independent, spatial relations. Descartes takes the spatial relations of bodies to be immediately grounded in the extension of bodies themselves. He thinks that the fact that one body is next to another body or between two other bodies is directly grounded in their extension and the extension of other bodies. Analogously, Du Châtelet holds that creatures have a kind of temporal ordering simply in virtue of their being successive per se. Their temporal order is not dependent upon their being related to an independent, absolute time, nor even to their standing in further, independent, temporal or causal relations. Du Châtelet takes the temporal relations of creatures to be immediately grounded in their successive being. She thinks that the fact that one creature exists immediately after or before another creature is directly grounded in their successive being and the successive being of other creatures, so that "there is no time without true, successive beings arranged in a continuous sequence; and there is time as soon as such beings exist" (1742, 102).

The third thesis is that the idea of time as an absolute, substantival being is false, natural, useful, and dangerous. It is false because it would violate the Principle of

[2] For further discussion of Du Châtelet's understanding of time, see Carus (forthcoming), Carus (2022), and Reichenberger (2021). See also the commentaries provided by Reichenberger (2018) and Hagengruber and Wells (n.d.).

Sufficient Reason. If time were an independent, homogenous backdrop to the things located in it, then there would be no reason for God to create sooner rather than later—a possibility that Du Châtelet rejects on the grounds that "one instant would have been chosen in preference to another to make this world actual without sufficient reason, which cannot be admitted" (1742, 96). She maintains that the idea of substantival time is nonetheless natural in that it arises in a psychologically spontaneous, perhaps even inevitable manner, when we think about successive beings and the temporal order that we abstract from them (1742, 97–99). That abstracted order, Du Châtelet claims, is also useful, and she tells us explicitly that our notion of absolute time "can have its use, when it concerns only the magnitude of the duration and comparisons of the duration of several beings together" (1742, 101). Finally, Du Châtelet cautions that although the false-if-natural idea of absolute time can be useful, it—like a hammer or a knife—can also be dangerous. For while the idea of absolute time can have its uses, it can also lead us astray in epistemology and metaphysics, and even ensnare us in heretical views concerning God's nature. Du Châtelet thus warns that although in geometry "the imaginary notion [of time] can easily be put in place of the real one … it is necessary to be cautious in metaphysics and in physics about making the same substitution, lest we fall into the difficulties of making duration an eternal being, and to which the attributes of God belong" (1742, 101).

With her understanding of time clarified, Du Châtelet suggests, at the end of Chapter 6 of the *Institutions*, that our idea of eternity is intimately related to our idea of ordinary time. In Sect. 116, she explains "how we acquire the idea of eternity:"

> To us the duration of all finite beings has a beginning and an end; now, if we remove from this idea that of the beginning, then the duration is eternity *a parte ante*; if we remove the end, the kind of duration is called eternity *a parte post*, and it is in this way that the soul of man is eternal. Finally, if we remove its beginning and its end from the idea that we have of the duration of finite beings, the duration will become *the eternity of God*, for only God can be eternal *a parte post*, and *a parte ante*, that is, having neither beginning nor end. (1742, 116)

Du Châtelet's account of how we acquire the idea of eternity naturally implies that she thinks that God is related to time in much the same way as creatures are related to time. God's duration seems to be distinguished from the duration of a flower or a farmer only in virtue of lacking all limits. Flowers have a beginning and an end in time. Farmers have a beginning but—given Christian doctrine—no end in time. Sect. 116 thus implies that God is a temporal being just like plants and people. If we assume the standard dichotomy between sempiternal and atemporal views, Sect. 116 must be taken to imply that Du Châtelet holds a sempiternal view of God's existence.

Sect. 116, so interpreted, would, however, stand in prima facie tension with Du Châtelet's remarks concerning God's eternity in Sect. 104 of the *Institutions*. Under a headline that reads "God is not in time and all succession is immeasurable for him," she writes:

> As for God, one cannot say that he is in time, for there is no succession in him since he is not capable of change. Thus, he is always the same without any variation; and just as he is outside the world, that is to say, as he is not linked with the beings whose union constitutes the world, he does not coexist with successive beings as creatures do; thus his duration cannot

> be measured by that of successive beings. For although God continues to exist during time, as time is only the order of the succession of beings, and this succession is immutable in relation to God, to whom all things with all their changes are present at once, God does not exist in time. God is at once all that he can be, whereas creatures can only successively achieve the states of which they are capable. (1742, 104)

If the passage from Sect. 116 might seem to suggest that Du Châtelet holds a sempiternal view of divine eternity, Sect. 104 might appear to imply that she holds an *atemporal* view of divine eternity. If we had to ascribe to her either a sempiternal or an atemporal view of divine eternity, Sect. 104—considered by itself—would clearly imply that she holds an atemporal view. For, on the standard dichotomy, that section naturally suggests that God is not eternal in the sense of enjoying a boundless temporal existence, but rather by not being temporal at all. Given the standard dichotomy, Sect. 104 conjures the analogy not of God's existence being like a boundless extended line, but rather of being like a point set off altogether from the extended timeline of creatures.

Of course, I think we should reject the standard dichotomy between sempiternal and atemporal views of eternity. If we do so, we can understand Sect. 104 as an affirmation of one half of the permissiveness indication. Given that, for Du Châtelet, time is "the order of successive things, insofar as they succeed each other, abstracting from any other internal quality than simple succession," it is easy to see why, on the permanent view, she should insist that there is a sense in which "God is not in time" (1742, 94, 104). For on the permanent view, God is not in any respect a "successive being," and so God can no more participate per se in the "order of successive things" than the letter B can belong per se to the order of integers, or the number 2 can belong per se to the order of the alphabet. But even setting aside the analytic point, I think we can see intuitively why, if time is understood as the order of successive beings, Du Châtelet should want to say that, in a sense, God is not in time. For it is clear from Sect. 116, that she holds God to be "always the same," to be immutable, to be "at once all that he can be." But such characteristics—all canonical—clearly imply that God cannot evolve or develop through time in the way that successive beings do. Successive beings—pure or otherwise—insofar as they are successive, temporal beings, are always—wholly or partially—coming into and passing out of existence. They, as Du Châtelet puts the point, "can only successively achieve the states of which they are capable" (1742, 104). Such temporal evolution, however, is entirely foreign to God understood as a permanent being. Holding that God is a purely permanent being, Du Châtelet would thus have naturally thought that God "is not in time," not just by the strict definition of time as "the order of successive things," but also intuitively in the sense that God is not at all a successive being, and so does not participate by his nature in time in the way that successive beings participate in time. For, "succession," as she puts it, "is an imperfection attached to finite" things alone (1742, 24).

If we reject the standard dichotomy between sempiternal and atemporal views of eternity, we can similarly understand Sect. 116 as an affirmation of the other half of the permissiveness indication. For, as we have noted, on the permanent conception of divine eternity, the thought that God does not participate per se in time does not require us to think of God as existing "outside of" or "apart from" time. Even a permanent being that does not participate in time by undergoing successive change per se may nonetheless endure through time. Rather than thinking of God as being like a point to the side of a line representing temporal succession, we may think of God as a point passing lengthwise through a line representing temporal succession. As we noted in the first section, the essences of creatures were once commonly held to be permanent beings. But it is certainly natural to think of the essences of creatures as existing in time even if we think of them as being permanent in the sense of being unchangeable, complete, and necessary. It should therefore seem equally natural to think of God as existing in time, or as co-existing in time, or at times, with creatures, even if we think of God as being permanent in the sense of being unchangeable, complete, and necessary. The supposition that Du Châtelet has such an understanding of divine eternity would thus not only allow us to understand why she thinks that, in a sense, "God does not exist in time" because "there is no succession in him, nor can any change happen to him," but also why she thinks that, in a sense, "God continues to exist during time" since even though "creatures can only successively achieve the states of which they are capable," nonetheless "God is at once all that he can be" (1742, 104). The next section will argue that Wolff holds a similar understanding of divine eternity and that his treatment of divine eternity in his *German Metaphysics* sheds additional light on aspects of Du Châtelet's discussion of God's eternity in Chap. 2 of her *Institutions.*

6.4 Wolff on Divine Eternity

Wolff agrees with Du Châtelet's three theses concerning time discussed above. First, like Du Châtelet, Wolff takes for granted a foundational notion of successive being. He implies, for example, that our idea of the simultaneity and the successiveness of things is prior to our idea of time, and that the latter may be understood in terms of the former (1977 [1730], 569–572; see also 578–579). Second, like Du Châtelet, Wolff conceives of time as the order of successive things, insofar as they succeed each other. In his *German Metaphysics*, he proposes, for example, that time is "nothing else than an order of that which follows one upon the other in such a way that if one takes one thing as the first, then another becomes the second, another the third, and so on" (2009 [1720], 94, see also 572). Third, like Du Châtelet, Wolff implies that absolute, substantival time is imaginary, natural, useful, and dangerous. He suggests, for example, that time considered as something independent of successive things is an "imaginary notion" (1977 [1730], 586). He suggests that such an imaginary notion of time comes to us naturally and is even useful to "mathematicians … regarding time only inasmuch as it may be measured" (1997 [1730], 581). He nonetheless warns of the perils of absolute time, cautioning that "One exposes oneself to the danger of erring when one, confusing an imaginary notion with the real one, considers time as a real entity existing outside successive things, and when one extends the imaginary notion beyond those limits within which it can be a substitute for the true one, and indeed considers time as an actual thing, which has flowed without beginning and will flow without end" (1977 [1730], 581).

Against the backdrop of his understanding of time, we see an early clue that Wolff inclines towards a permanent view of divine eternity in his discussion of essences in his *German Metaphysics* (see, 2009 [1720], 33–42; 1977 [1730], 143–156). There, in keeping with the second, grounding indication mentioned above, Wolff suggests that essences are eternal *because* they are necessary, that their being eternal follows from their mode of existence rather than from their relation—atemporal or sempiternal—to time. Wolff begins his discussion by explaining that an essence is that "in which the ground is to be found for the rest of what pertains to a given thing," adding that "Whoever cognizes the essence of a thing can indicate the ground for everything that pertains to it" (2009 [1720], 33). He then insists that since "the essence of a thing consists in the fact of its being possible in a certain way and manner" it follows that the "essence of a thing is necessary" (2009 [1720], 38). Next—and most importantly for our purposes—Wolff goes on to explicitly declare that "That which is necessary is eternal," thereby indicating that he sees the eternality of essences as following from their necessity rather than from their bearing any special relationship to time (2009 [1720], 39). In keeping with the grounding indication, essences, for Wolff, are eternal not because they stand outside of time, nor because they exist at all times, but because they are necessary, that is, because of the nature of their existence. Wolff's understanding of the grounds of eternality gives us a first clue that he is committed to a permanent, rather than to a specifically atemporal or a sempiternal view of divine eternity.

The initial impression that Wolff has a permanent view of essences is strengthened by his commitment—in keeping with the permissiveness indication—to essences being both, in a sense, "out of time," and, in a sense, existing at all times. In discussing essences, Wolff insists that "That which is necessary is also unchangeable," adding explicitly, if a bit redundantly, that "On account of the fact that the essence of a thing is necessary (§38), it is also unchangeable" (2009 [1720], 42; compare 2009 [1720], 46). Given, however, that time, for Wolff, is "nothing else than an order of that which follows one upon the other in such a way that if one takes one thing as the first, then another becomes the second, another the third, and so on," it follows straightforwardly that there is a sense in which essences cannot participate per se in the order of successive being, and therefore there is a sense in which essences are not in time (2009 [1720], 94). Nonetheless, and conversely as it were, since Wolff also takes essences to be aspects of both composite and simple creatures, he must also allow that there must be a sense in which they—as aspects of creatures—exist in time (see, 1977 [1720], 542). But in whatever sense essences may be said to exist in time, they must also be said to exist at all times, since, according to Wolff, "That which is necessary … can have neither beginning nor end. For, if something is necessary, then it is impossible for it not to be" (2009 [1720], 39). Wolff's discussion of essences in his *German Metaphysics* thus suggests that his understanding of the eternality of essences is not only consistent with the grounding indication but also with the permissiveness indication, that is, with essences being both, in a sense, not in time—they can't participate per se in the order of successive being – and, in a sense, sempiternal—they must endure through time without beginning or end. Wolff thus appears to hold, in general, a permanent view of essences.

Given that God was commonly held to be identical to his essence, Wolff's general treatment of essences ought to raise the suspicion that he also has a permanent view of divine eternity. That suspicion is confirmed, I think, in Chapter 6 of Wolff's *German Metaphysics*. There, in keeping with both the grounding indication and—as it were—the second half of the permissiveness indication, Wolff suggests that God's existing at all times follows from his permanent, necessary existence. In Sect. 928, Wolff opens by arguing that there must be a necessary being since we exist, and "Everything that is has its sufficient ground why it is rather than is not," and such a ground can ultimately be located only in the existence of a necessary being (2009 [1720], 928). Wolff then argues—apparently by definition—that that necessary thing must also be an independent thing since "That thing which contains the reason of its own actuality and therefore is such that it is impossible for it not to be, is called an *independent being*" (2009 [1720], 929). Wolff next suggests that a necessary, independent being must also be eternal in the sense of not having a temporal beginning or end. Of the necessary, independent being established in Sect. 928, he writes "It is eternal," explaining that "That which is necessary can have neither a beginning nor an end, but is rather eternal (39)," adding specifically, if again somewhat redundantly, that "the independent being is necessary (929), so it can have neither beginning nor end, but is instead eternal" (2009 [1720], 931). Sections 928–931 of Wolff's *German Metaphysics* thus suggest that Wolff, in keeping with the grounding indication, sees God's eternality as following from his essence, and, in keeping with sempiternal half of the permissiveness indication, takes there to be a sense in which God may be said to exist at all times without beginning or end (just as Du Châtelet does in Sect. 116 of her *Institutions*).

Sections 928–931 of Wolff's *German Metaphysics* might seem to leave open the possibility that, for Wolff, God is eternal simply by being sempiternal, that is, simply by existing at all times. It is significant therefore that roughly a hundred sections later, Wolff explicitly insists—in keeping with the first half of the permissiveness indication—that there is a sense in which God does not exist in time at all. In Sect. 1072, Wolff begins by establishing to his own satisfaction that "God is infinite." He argues that "Because God represents everything that is possible all at once and indeed in all distinctness (§955), and all that can pertain to Him is contained in this representation (§1067), so God is everything that He can be, all at once, and therefore without any limitations, that is, He is infinite (§109)" (2009 [1720], 1072). From God's being infinite, Wolff proposes to deduce that God is also unchangeable, arguing that "it further becomes clear that God is unchangeable. For if something were supposed to change in God, then He would be something that He had not been before, and accordingly He would not be everything at once, and consequently not infinite, which is contrary to what was just proven (§1072)" (2009 [1720], 1073). But from the fact that God is unchangeable, it follows that God does not participate per se in the order of successive beings, that is, it follows that God does not participate per se in the order of time. Wolff draws this conclusion explicitly, writing that "because God is everything at once, so nothing precedes, and nothing follows after (§1072). As a result of this, since time consists in the order of things that follow one upon the other (§94), God is not in time." (2009 [1720], 1074). If Sects. 928–931 confirm the second half of the permissiveness indication by suggesting that there is a sense in which God exists sempiternally, Sects. 1072–1074 confirm the first half of the permissiveness indication by insisting that there is nonetheless a sense in which God does not exist in time at all. Taken together, these two stretches of texts—Sections 928–931 and Sects. 1072–1074—indicate that Wolff holds neither a sempiternal nor an atemporal view of divine eternity. Taken together, they imply that Wolff, like Du Châtelet, is committed to permanent view of divine eternity.

Wolff's discussion of divine eternity in his *German Metaphysics* helps to shed additional light on Du Châtelet's discussion of divine eternity in Chap. 2 of her *Institutions.* Like Wolff, Du Châtelet begins her discussion of God in Chap. 2 by arguing that a necessary being must exist.[3] In the course of that argument, Du Châtelet not only insists that a necessary being must exist "from all eternity," she also suggests that God's necessary, eternal existence must be contrasted with the contingent, successive existence of creatures. She thus writes that "All that is around us is born and perishes successively; nothing enjoys a necessary state, everything is successive, and we ourselves succeed each other … for this is what distinguishes a contingent being from a necessary being" (1742, 19.4). The same contrast between God's necessary, eternal existence and the successive existence of creatures crops up in the 1740 edition of the *Institutions* when she argues that "Neither the World nor Our Soul Can Be a Necessary Being" (1740, 21). Under that heading, she argues that "The world we see cannot be the necessary Being, for it is composed of parts and *there is a continual succession in it*, which is absolutely contradictory to the attributes … shown to belong to a necessary Being" (1740, 21, emphasis added). She likewise suggests that neither matter nor the elements of matter can be the necessary Being" for the same reason—that is, because "there is a continual succession in it" (1740, 21). Finally, she argues that even "our soul" cannot be "this necessary Being, for its [i.e., the soul's] perceptions, changing continually, it is in perceptual variation, but the necessary Being cannot vary." She concludes that because it has a successive existence "our soul is not the necessary Being" (1740, 21).

Du Châtelet's emphasis on the contrast between the necessary, eternal existence of God and the contingent, successive existence of creatures coheres well with the suggestion that she—like Wolff—holds a permanent view of divine eternity. For, in keeping with the grounding criterion, Du Châtelet clearly maintains that divine eternity is grounded in God's necessary existence. God's eternality follows from his being a necessary being, not from his existing outside of time nor from his existing at all times. In keeping with the first half of the permissiveness criterion, Du Châtelet also maintains explicitly in Chap. 2 that God exists at all times. Under a heading that reads *He is Eternal*, she insists that it follows from "the necessity for its existence" that God is eternal in the sense of having no beginning or end in time. God can have no beginning in time, she explains, "for if the necessary

[3] For discussion of Du Châtelet's argument for the existence of God and its historical context, see Lascano (2011).

Being had begun, it would have to have acted before being, in order to produce itself, which is absurd, or something must have produced it, which is contrary to the definition of the necessary Being" (1742, 20). Likewise, God can have no end in time "because the sufficient reason for its existence residing in itself, it can never abandon it; furthermore, it implies a contradiction that a necessary being not exist (N.3): but what implies a contradiction is impossible. It is therefore impossible for the necessary Being to cease existing, just as it is impossible for 3 times 3 to make 8" (1742, 20). Like Wolff, in his discussion of God's existence, Du Châtelet, in her discussion of God's existence, commits herself not only to the view that God is eternal because his existence is necessary, but also to the view that there is a sense in which God's existence is sempiternal.

In her discussion of God's existence in Chap. 2 of the *Institutions*, Du Châtelet does not explicitly affirm the second clause of the permissiveness indication. That is to say in Chap. 2 of the *Institutions*, Du Châtelet does not explicitly state that there is a sense in which God may be said to be "not in time." Nonetheless in Sect. 20 of the *Institutions*, she does argue that God must be immutable, writing that "for if it [that is, God] changed, it would no longer be what it was, and consequently it could not have existed necessarily" (1742). After contrasting God's immutable existence with the successive existence of creatures, Du Châtelet concludes that "there cannot be change or succession in a necessary Being" (1740, 20). Thus, although she doesn't explicitly affirm the second half of the permissiveness indication in Chap. 2 of the *Institutions*, Du Châtelet's insistence that God is immutable when paired with her understanding of time does entail that there is a sense in which God is not in time. For if time is the order of successive being, and God is not per se a successive being, then God cannot participate per se in time. For Du Châtelet, God's being eternal in the sense of God's having no beginning nor end in time—that is, God's existing sempiternally—is consistent with God's also being eternal in the sense of God's not participating per se in the order of time—that is, with God's existing atemporally. Du Châtelet's discussion of God's existence in Chap. 2 of the *Institutions* thus again dovetails both with Wolff's discussion of divine eternity in his *German Metaphysics* as well as with the interpretation offered above of Du Châtelet's discussion of divine eternity in Chapter 6 of the *Institutions.*

6.5 Leibniz on Divine Eternity

The relationship between the views of Leibniz and Du Châtelet on divine eternity is less clearcut than the relationship between the views of Wolff and Du Châtelet. On the one side, Leibniz agrees with much of Du Châtelet's understanding of time. Like Du Châtelet, he maintains that time is essentially relational. In his Third Letter to Clarke, he writes, for example, that "I have said more than once that I hold space to be something purely relative, as time is—that I hold it to be an order of coexistences, as time is an order of successions" (Leibniz & Clarke, 2000, [3:4] 14; see also [5:33] 42). He likewise agrees with Du Châtelet that time is in some sense ideal or abstract.

He tells us, for example, that "the parts of time or place considered in themselves are ideal things" and he claims that anyone who follows his arguments "will easily apprehend that time can only be an ideal thing," going on to note that "the analogy between time and space will easily make it appear that the one is as merely ideal as the other" (Leibniz & Clarke, 2000, [5:27] 41; 2000, [5:49] 48). Finally, Leibniz and Du Châtelet agree in their rejection of absolute time. Like Du Châtelet, Leibniz insists that Newton's and Clarke's substantival view is both false and dangerous. Absolute time, like absolute space, he insists, must be rejected as a "chimerical supposition" (Leibniz & Clarke, 2000, [3:5] 15). Its postulation would force us to suppose – contrary to both reason and piety—that "God has done something concerning which it is not possible that there should be a reason why he did it so and not otherwise" (Leibniz & Clarke, 2000, [3:6] 15).

On the other side, however, Leibniz departs from Du Châtelet's understanding of time in at least two important respects. First, although both Leibniz and Du Châtelet agree that time is ideal, they have very different views concerning what it means for time to be ideal. Leibniz is essentially a Platonist. He insists that time is ideal insofar as it is a necessary, immutable structure in the divine intellect. Like a Platonic form, ideal time in this sense is not dependent upon the actual order of successive beings, but rather serves as a necessary parameter on any possible order of successive beings (Leibniz & Clarke, 2000, [3:4] 14; see also 1875–1890, 2:379). It is for precisely this reason that Leibniz insists that time is not merely an order of actual created entities but rather an order of all possible created entities (Leibniz & Clarke, 2000, [3:4] 14; [5:104] 60–61; see also Leibniz 1875–1890, 2:379). Second, although Leibniz and Du Châtelet both insist that actual time is (partially) grounded in the successiveness of created beings, Leibniz aims to ultimately ground the successiveness of created beings in something further, in the law of non-contradiction and causal-explanatory order (Leibniz 2001, 269). Although the details of the project are subtle, the contrast with Du Châtelet is clear. Where Du Châtelet sees time as being (partially) grounded in something that is fundamentally temporally ordered—much like Descartes sees space as being grounded in something that is fundamentally spatially ordered—Leibniz wishes to reduce actual temporal order to something more fundamental, something like logical inconsistency and causal or explanatory order (just as he wishes to reduce actual spatial order to something more fundamental, namely, logical consistency and points of view) (Leibniz 2006, 6.4.568; 6.4.629; for discussion, see Arthur 2021, Chapter 1).

There are tantalizing indications that in his early years Leibniz held a permanent view of divine eternity not unlike Du Châtelet's view. In keeping with the grounding indication, in an early piece, he writes that "eternity is … the necessity of existing, which does not express succession, duration or divisibility" (2006, 6.3.392). In another piece, he tells us that "absolute existence is eternity or [*sive*] necessity" (2006, 6.3.396; see also Leibniz 1875–1890, 6:439). Even more enticingly, however, in still another early text, Leibniz seems to be in agreement with the permissiveness indication. In what has been entitled, *A Notable Observation on Infinity*, he writes that (i) "Eternity, if it is conceived as something which is homogeneous with time, will be an unbounded time; if as an attribute of something eternal, it will be duration through an unbounded time. But [(ii)] the true origin and inmost nature of eternity is the very necessity of existing, which does not in itself indicate succession, even if it should happen that what is eternal should coexist with all things" (2006, 6.3.484). Given the obscurity of the topic, one could hardly hope for a clearer affirmation of the permissiveness indication. The first clause—clause (i)—suggests that we can conceive of eternity as either an unbounded time or an unbounded enduring through time, but, in either case, as a kind of sempiternity. The second clause—clause (ii)—however suggests that even if an eternal thing should endure through time, and "happen to coexist with all things," by its inmost nature it does not participate per se in the temporal order. If clause (i) suggests that divine eternity may be understood as sempiternity, clause (ii) suggests that it may be understood as atemporality, and the two clauses together, in keeping with the permissiveness indication, clearly suggest that the early Leibniz at least flirted with a permanent conception of divine eternity like the one that Du Châtelet would later embrace.

The mature view that Leibniz offers in opposition to Clarke, however, appears to be different. In their famous correspondence and elsewhere, Clarke defends a complex, not obviously consistent, but nonetheless clearly sempiternal, view of divine eternity. We might think of it as involving three main commitments. First, Clarke maintains that God's thoughts are successive (see Vailati, 1997, 21, 34, 36, 59). If they were not, Clarke insists, God could not "vary his Will, nor diversify his Works, nor act successively, nor govern the World nor indeed have any Power to will or do anything at all" (Clarke, 1738, 3:896). This first commitment implies, of course, that, for Clarke, God participates in the temporal order per se, that is, that God is a temporal being. Second, Clarke maintains that God is present at all times not merely by dint of his power, but more robustly by being substantially present at all times. Opposing Leibniz's suggestion that God's "presence is manifested [merely] by his immediate operation" (Leibniz & Clarke, 2000, [3:12] 16–17), Clarke responds that "God, being omnipresent, is really present to every thing, essentially and substantially. His presence manifests itself indeed by its operation, but it could not operate if it was not there" (Leibniz & Clarke, 2000, [3:12] 20; see Brown (2016), 42–46 for discussion). This second commitment thickens the sense in which God is temporal. By Clarke's lights, God not only has temporally successive thoughts, he is literally present by his very substance at all passing times. Third, Clarke maintains that although different senses of eternal can and should be distinguished, God exists in the "last and highest and most absolutely perfect Sense of the words, eternal and everlasting" (Clarke, 1738, 1:21). God, according to Clarke, exists with a "Duration of inexhaustible and never-failing permanency, both without Beginning, and without End" (Clarke, 1738, 1:21). While he grants that there is a technical sense in which God does not exist in time because time is an attribute dependent upon God rather than the other way around, Clarke nonetheless clearly embraces a sempiternal view of divine eternity, a view on which God exists—Clarke's technical caveat aside—successively through time without beginning or end.

Clarke's sempiternalism provides a helpful backdrop for appreciating the atemporalism of Leibniz's mature writings. In contrast to Clarke's first commitment,

Leibniz flatly rejects the suggestion that God's thought is successive. He writes in the *Theodicy*:

> [T]he divine understanding has no need of time for seeing the connexion of things. All trains of reasoning are in God in a transcendent form, and they preserve an order amongst them in his understanding, as well as in ours: but with him it is only an order and a priority of nature, whereas with us there is a priority of time. (Leibniz 1875–1890, [§192], 230; cf. Du Châtelet, 1742, 24)

Almost 200 sections later, Leibniz makes the same point again suggesting that while "the decrees of God have an order among themselves" they are "found in him in a transcendent way" so that "one sees that there is in the propositions or truths a natural order; but there is no order of time or interval to cause him to advance in knowledge and pass from premises to the conclusion" (Leibniz 1875–1890, [§389], 346). As Michael Futch notes, these passages are not strictly inconsistent with God's being temporal (Futch, 2008, 193). One could maintain that there is no temporal order to God's understanding and yet maintain that God endures through time. Nonetheless, in context, Leibniz's rejection of the successiveness of God's thoughts and will provides a sharp contrast with Clarke and at the least a first clue that the mature Leibniz holds an atemporal view of divine eternity.

A second clue is provided by Leibniz's account of divine omnipresence. As we saw just above, Clarke, following the principle that God can only act where he exists, insists that God must be substantially present everywhere and everywhen. Leibniz explicitly rejects Clarke's suggestion that God is substantially present everywhere. He is adamant that if God were substantially present everywhere, God would have parts, and would thus be divisible – a conclusion that according to him contradicts divine simplicity (Leibniz & Clarke, 2000, [3:3] 14; [5:42–43] 45). God, according to Leibniz, thus exists in all places not substantially, but rather by his "will" and "his immediate operations" (1996, 155). Leibniz's thought that substantial omnipresence must entail divine parthood is not, of course, incontrovertible. There are messy questions in this area about what sense of parthood and divisibility are at stake in the issue. If distinguishable regions of space are necessary and necessarily connected, are they parts, and if they are parts, are they parts in a sense that offends divine simplicity? But that is not the important point for our present concerns. The important point for our present concerns is that whatever the force of Leibniz's argument against God's being substantially present everywhere, the same line of thought seems to apply equally well against God's be substantially present everywhen. Someone accepting the conditional "If God were substantially present everywhere, then God would have spatial parts," should also accept the conditional "If God were substantially present everywhen, then God would have temporal parts." The implication, of course, is clear. Leibniz's argument against God's being substantially present everywhere gives him an equally good argument against God's being substantially present everywhen. Put in different words, one of Leibniz's chief complaints against Clarke's understanding of God's relationship to space, gives him a powerful reason for concluding that God is not himself substantially present in time. That is, for concluding that God is, in a sense, atemporal.

Finally, a third clue in support of Leibniz's mature embrace of an atemporal understanding of divine eternity is provided by his suggestion that God's "immensity and eternity" are "more transcendent" than even endless creaturely duration. In a passage to which others have called attention (see, in particular, Futch, 2008, 192), Leibniz writes:

> It is true that the immensity and eternity of God would subsist though there were no creatures, but those attributes would have no dependence either on times or places. If there were no creatures, there would be neither time nor place, and consequently no actual space. The immensity of God is independent of space as his eternity is independent of time. These attributes signify only <with regard to these two orders of things> that God would be present and coexistent with all the things that should exist. And therefore I do not admit what is here advanced, that if God existed alone, there would be time and space as there is now, whereas then, in my opinion, they would be only in the ideas of God as mere possibilities. The immensity and eternity of God are things more transcendent than the duration and extension of creatures, not only with respect to the greatness, but also to the nature of the things. Those divine attributes do not imply the supposition of things extrinsic to God, such as are actual places and times. These truths have been sufficiently acknowledged by theologians and philosophers. (Leibniz & Clarke, 2000, [5:106] 61)

There is a great deal going on in this rich passage, but the most important point for our purposes is simply that Leibniz is insisting that God's eternity is independent of actual time. If there were no creatures, there would be no actual time, but God's eternity would be unaffected. For the mature Leibniz, the "immensity and eternity of God are things more transcendent than the duration and extension of creatures, not only with respect to the greatness, but also to the nature of the things." But this clearly implies that God's eternity is not to be identified with sempiternity. If God's eternity is independent of time, then it cannot be identified with endless existence in time. As Futch puts the point, "If God is eternal but can exist without time, then God's eternity cannot be understood in temporal terms. Put differently, the possibility of an eternal God existing without time implies that eternity is not sempiternity" (Futch, 2008, 192). It also implies, we should add, that Leibniz is not thinking, in this passage at least, of God's eternity in terms of permanent existence. For as we have seen, it is an integral part of that conception of divine eternity, that God can be thought of not only as existing (in a sense) atemporally but also as existing (in a sense) sempiternally. Insofar as the quoted passage rules out one member of that conjunction, it implies that Leibniz is thinking of God's existence as being neither sempiternal nor as being permanent. Leibniz's mature view—in contrast to both Du Châtelet's and Wolff's—is that God's existence is fully atemporal.

6.6 Conclusion

In this essay, I have argued for two core conclusions. First, Du Châtelet is best read as holding a permanent view of divine eternity. On such a reading, Du Châtelet belongs to a long tradition of thinkers who thought of God's eternity as being most directly grounded in the nature of his necessary, immutable, complete existence.

Because they saw God's eternity as being immediately grounded in a special kind of existence, they could, and did, allow that there is a sense in which God may be said to be in time—he endures through time without beginning or end—and a sense in which God may be said to be out of time—in contrast to creatures like ourselves, he does not participate in the temporal order of the world per se. Such a reading allows us to make sense both of Du Châtelet's otherwise apparently conflicting statements about divine eternity in Chapter 6 of the *Institutions* as well as her remarks about God's nature in Chap. 2 of the *Institutions*. Second, although Du Châtelet, Wolff, and Leibniz all insist that time is relational and that the substantivalism of Newton and Clarke is a dangerous mistake, Du Châtelet's views on divine eternity are ultimately closer to Wolff's views than to Leibniz's mature views. Du Châtelet and Wolff share a permanent view of divine eternity and embrace its implications concerning God's nature and relation to time. The young Leibniz may well have shared their views, but the mature Leibniz of the *Correspondence*, the *New Essays*, and the *Theodicy*, appears to embrace a more thoroughly atemporal view of God's relationship to time. Whereas for Du Châtelet and Wolff, God endures through time without participating per se in the order of successive being, for the mature Leibniz, God sits outside the temporal order of the natural world altogether.

References

Aristotle. (1983). *Physics*. E. Hussey (Trans.). 1st ed. Clarendon Press.

Arthur, R. T. W. (2021). *Leibniz on time, space, and relativity*. Oxford University Press.

Brown, G. (2016). Did Samuel Clarke Really Disavow action at a distance in his correspondence with Leibniz? Newton, Clarke, and Bentley on gravitation and action at a distance. *Studies in History and Philosophy of Science Part A, 60*, 38–47.

Carus, C. (2022). Du Châtelet's contribution to the concept of time. In R. Hagengruber (Ed.), *Epoque Emilienne: Philosophy and science in the age of Emilie Du Châtelet* (pp. 113–128). Springer.

Carus, C. (Forthcoming). Comparing Du Châtelet's and Kant's epistemological approach to space and time. In R. Hagengruber & A. Wells (Eds.), *Émilie Du Châtelet and Kant*. Springer.

Clarke, S. (1738). *The works of Samuel Clarke*. Harvard College Library Pamphlet Digitization Program. Printed for J. and P. Knapton. http://nrs.harvard.edu/urn-3:FHCL:10738956

Deng, N. (2023). Eternity in Christian thought. In *The Stanford encyclopedia of philosophy*. In E. N. Zalta & F. Uri Nodelman (Eds.), Metaphysics Research Lab, Stanford University. https://plato.stanford.edu/archives/fall2023/entries/eternity/

Du Châtelet, É. (1740). *Institutions de physique*. Prault fils.

Du Châtelet, É. (1742). *Institutions physiques*. Aux dépens de la Compagnie.

Du Châtelet, É. (2009). *Selected philosophical and scientific writings*. In J. P. Zinsser (Ed.). (I. Bour & J. P. Zinsser, Trans.). Illustrated ed. University of Chicago Press.

Du Châtelet, É. (2018). *Foundations of physics*. (Trans.) K. Brading, et al. https://www.kbrading.org/translations

Dyck, C. W. (Ed.). (2019). *Early modern German philosophy (1690–1750)*. Oxford University Press.

Futch, M. (2008). *Leibniz's metaphysics of time and space*. Springer Science & Business Media.

Hagengruber, R., & Wells, A. (n.d.). Chapter Six of Du Châtelet's Institutions de Physique. Historyofwomenphilosophers.Org. https://historyofwomenphilosophers.org/journals/journal/chapter-six-of-du-chatelets-institutions-de-physique/. Accessed March 1, 2024

Lascano, M. P. (2011). Emilie Du Châtelet on the existence and nature of god: An examination of her arguments in light of their sources. *British Journal for the History of Philosophy, 19*(4), 741–758.

Leibniz, G. W., & Clarke, S. (2000). *Leibniz and Clarke: correspondence* (UK ed.). In R. Ariew (Ed.). Hackett Publishing Company, Inc.

Leibniz, G. W. (1875–1890). *Die Philosophischen Schriften Von Gottfried Wilhelm Leibniz* (Vol. 7). In C. I. Gerhardt (Ed.). Weidmann.

Leibniz, G. W. (1996). *New essays on human understanding. Cambridge texts in the history of philosophy.* Cambridge [Eng.]. Cambridge University Press.

Leibniz, G. W. (2006). *Philosophische Schriften.* Verkleinerter Reprint der Erstausg. Akademie Verlag.

Leibniz, G. W. (2001). *The labyrinth of the continuum: Writings on the continuum problem, 1672–1686.* In R. T. W. Arthur (Trans.). Yale University Press.

McDonough, J. K. (Forthcoming). Time for Émilie Du Châtelet.

Melamed, Y. Y. (Ed.) (2016). *Eternity: A history. Oxford philosophical concepts.* Oxford University Press. https://doi.org/10.1093/acprof:oso/9780199781874.001.0001

Pasnau, R. (2011). On existing all at once. In C. Tapp (Ed.), *God, eternity, and time.* Ashgate.

Pasnau, R. (2013). *Metaphysical themes 1274–1671.* UK ed. Oxford University Press.

Reichenberger, A. (2018). *Du Châtelet's Institutions of Physics. An Online Reading Guide.* Center for the History of Women Philosophers and Scientist HWPS, Paderborn University.

Reichenberger, A. (2021). Émilie Du Châtelet on space and time. *Revue D'histoire des Sciences Tome, 74*(2), 331–355. https://doi.org/10.3917/rhs.742.0331

Vailati, E. (1997). *Leibniz & Clarke: A study of their correspondence.* Oxford University Press.

Wolff, C. (1977). *Philosophia prima, sive, Ontologia.* Edited by J Ecole and Charles Corr. 2. Nachdruckaufl. Gesammelte Werke/Christian Wolff. Hildesheim: G. Olms. http://catalog.hathitrust.org/api/volumes/oclc/46362536.html

Wolff, C. (2009). *Vernünfftige Gedancken: von Gott, der Welt und der Seele des Menschen, auch allen Dingen überhaupt 1 …* Edited by Charles A. Corr and Jean Ecole. Gesammelte Werke/ Christian Wolff. Hildesheim: Olms.

Chapter 7
Du Châtelet's Rejection of Leibniz's World Apart Doctrine

Fatema Amijee

Leibniz famously writes that "each substance is like a world apart, independent of all other things, except for God".[1] According to the standard reading of Emilie Du Châtelet's metaphysics as laid out in her *Institutions de Physique*—a reading that has come under increasing pressure in recent commentaries—Du Châtelet's metaphysics is inherited from Leibniz.[2] Yet, in a remarkable departure from Leibniz's 'world apart doctrine', Du Châtelet endorses the view that created substances stand in transeunt causal relations to one another, or so I will argue. To be sure, this apparent endorsement sits oddly with Du Châtelet's own claim in her preface that her goal in the first chapters is to explain "the principal opinions of M. Leibniz on metaphysics".[3] But by her own admission, Du Châtelet's knowledge of Leibniz's metaphysics came to her filtered through other Leibnizians, including Wolff, and so may be incomplete as well as inaccurate.[4] My goal in this paper is not primarily

[1] 'Discourse on Metaphysics', §14/AG 47.

[2] Cf. Lascano (2011) and Stan (2018). Lascano (2011, 742), for instance, writes that "[m]any commentators have written about the *Institutions*, but the metaphysical sections are almost always described as a mere retelling—and somewhat inaccurate one at that—of Leibniz."

[3] See IP, Preface (XII). I cite the 1740 edition of the *Institutions* (abbreviated 'IP') by section, and use translations from chapters of the *Institutions* translated by Brading et al. (https://www.kbrading.org/translations) and from Zinsser and Bour (2009).

[4] It is worth noting that Du Châtelet does herself distinguish between the views of Leibniz and those of Wolff, writing the following at IP §119: "Few people in France know anything of this opinion of M. Leibniz's but the word, *monads*; the books of the famous Wolff, in which he explains so clearly and eloquently M. Leibniz's system, which, in his hands, *took a totally new form*, have not yet been translated into our language" (emphasis added). It is also significant that Du Châtelet does not herself use the term "monads" for her elements or simple substances, which suggests that she may consciously be putting some distance between her views and those of Leibniz's.

F. Amijee (✉)
University of British Columbia, Vancouver, Canada
e-mail: famijee@gmail.com

C. Carus and J. K. McDonough (eds.), *Émilie Du Châtelet in Relation to Leibniz and Wolff*, Women in the History of Philosophy and Sciences 23,
https://doi.org/10.1007/978-3-031-80541-7_7

to examine the extent to which Du Châtelet's stated views about substance come apart from Leibniz's views and trace other sources of influence, but to address what is in the first instance a philosophical question: does Du Châtelet's commitment to creaturely transeunt causation between substances (i.e., to causation whereby one created substance productively or efficiently causes an accident in another created substance), and consequently her rejection of the causal version of Leibniz's 'world apart' doctrine, flow from or inform some of her other metaphysical commitments? I will argue that it does. In particular, I will show that Du Châtelet's commitment to creaturely transeunt intersubstantial causation is philosophically undergirded by her commitment to a conception of a natural or non-divine sufficient reason as a reason that is in principle accessible by some finite mind or other, as well as by her causal account of intentionality.

I proceed as follows. In §1, I discuss Leibniz's world apart doctrine, with a particular focus on the causal independence of substance. In §2, I argue that Du Châtelet, unlike Leibniz, embraces the possibility of transeunt causation both amongst simple substances which ground bodies and between souls and simple substances that ground bodies. In §3, I trace Du Châtelet's commitment to transeunt intersubstantial causation to her commitment to the view that any natural (i.e., non-divine) sufficient reason is in principle accessible by us (that is, by a finite mind or other), and to a causal theory of intentionality. Finally, in §4, I put Du Châtelet's characterizations of sufficient reason and of substance into sharp relief by contrasting them with Leibniz's characterizations of sufficient reason and substance, both of which inform Leibniz's commitments about causation.

7.1 Causal Independence in Leibniz

As stated above, Leibniz is committed to the doctrine that each substance is independent of all other things, except for God. If a substance is independent of *all* other created entities, it is *a fortiori* independent of all other created substances.[5] Leibniz endorses the independence of substance along several distinct dimensions. Among other senses, Leibniz holds that substances are modally independent (where something is modally independent just in case it does not require the existence of any other created substance), conceptually independent (where something is conceptually independent just in case its concept does not contain the concept of any other

[5] If independence is meant to function as a (partial) *criterion* for substance for Leibniz, then an elucidation of the relevant notion of independence cannot itself involve the notion of a created substance. However, if Leibniz's world apart doctrine is not meant to function as a *criterion* for substance, then independence can be characterized in terms of independence from other created substances [see Harmer (2016) who distinguishes three varieties of independence in Leibniz—causal, ontological and phenomenal—and characterizes them in this way].

created substance), and causally independent (where something is causally independent just in case it does not causally depend on any other created substance).[6] For the purposes of this paper, I focus exclusively on Leibniz's commitment to the causal independence of substance, for it is the independence of substance in this sense that (I will argue) Du Châtelet most directly rejects.

Leibniz puts forward two main objections to the possibility of transeunt causation between created substances (henceforth sometimes just 'intersubstantial causation').[7] According to the first, intersubstantial causation requires the transfer of accidents from one substance to another; but accidents, as dependent entities, cannot detach themselves from substances, and so intersubstantial causation is unintelligible.[8] For example, he writes in the *Monadology*:

> There is also no way of explaining how a monad can be altered or changed internally by some other creature, since one cannot transpose anything in it...The monads have no windows through which something can enter or leave. Accidents cannot be detached, nor can they go about outside of substances, as the sensible species of the Scholastics once did. Thus, neither substance nor accident can enter a monad from without.[9]

See also Leibniz's letter to de Volder of 20 June 1703/AG 176:

> Properly speaking, I don't admit the action of substances on one another, since there appears to be no way for one monad to flow into another.[10]

And another of Leibniz's letters to de Volder (written in 1704 or 1705) (emphasis added):

> It is necessary that these simple substances exist everywhere and that they be self-sufficient (with respect to one another), since an influx of one into another is *unintelligible*, beyond the fact that such an influx is something placed in things for no purpose, and established by no argument.[11]

---

[6] Other varieties of independence may also play a role for Leibniz, such as mereological or constitutional independence (x mereologically/constitutionally independent just in case x there is nothing that is a part of x or constitutes x) and predicational independence (x is predicationally independent just in case there is nothing that it inheres in or is predicated of). Harmer (2016) isolates and discusses yet another notion of independence: phenomenal independence. He writes that "every (created) substance is phenomenally independent of all the others, i.e., the truth of a substance's phenomena (roughly, what it sees) does not require the existence of any other substances except God." (39). I bracket these other senses for present purposes.

[7] Cf. Watkins (1995, 297).

[8] If causation is construed more broadly such that counterfactual dependence suffices for the obtaining of a causal relation between two substances, one could say that Leibniz too is committed to causation between created substances, for Leibniz's substances exist in the kind of preestablished harmony that would satisfy counterfactual dependence between created substances. But the sense of causation relevant to Leibniz's denial of causation between substances, and thus the sense relevant to this paper, is much narrower: it refers to transeunt (or productive or efficient) causation.

[9] §7, G VI 607/AG 214.

[10] G II 241/AG 176.

[11] G II 275/AG 181.

According to Leibniz's second objection, transeunt intersubstantial causation involves a violation of the law of conservation of motion. As Leibniz writes in §80 of the *Monadology*:

> Descartes recognized that souls cannot impart a force to bodies because there is always the same quantity of force in matter. However, he thought that the soul could change the direction of bodies. But that is because the law of nature, which also affirms the conservation of the same total direction in matter, was not known at that time. If he had known it, he would have hit upon my system of pre-established harmony.[12]

Yet, even as Leibniz rejects the intelligibility of transeunt causal relations between created substances, he does not reject all causal activity when it comes to created substances. Leibniz endorses *intra*substantial causation, i.e., causation within a substance. Indeed, it is essential for Leibniz that substances are causally active: in his *Principles of Nature and Grace, Based on Reason*, Leibniz defines a substance as "a being capable of action".[13] Nor does Leibniz reject all transeunt (i.e., efficient or productive) causation *between* substances: God, for Leibniz, creates the world and acts on it, albeit not through a transfer of accidents from God to created substances.[14] I will return to Leibniz's rejection of intersubstantial causation, and its relationship to his account of substance, in §4.

7.2 Du Châtelet on Causal Interaction

I will argue that, unlike Leibniz, Du Châtelet is committed to transeunt causation between created substances. In what follows, I will make a textual case for this commitment, and then briefly address whether her divergence from Leibniz on the question of intersubstantial causation may be traced to the views of Christian Wolff. Let us begin, however, with a preliminary question: what sorts of entities qualify as substances for Du Châtelet?

Unlike Leibniz, who at times explicitly endorses a commitment to composite substances, and at times even commits himself simultaneously to both substances that are corporeal and those that are immaterial, Du Châtelet writes at IP §127 that "simple beings are the *real* substances, beings that are durable and susceptible to the modifications which their internal force produces" and at IP §128 that "the *real* substances (that is to say), simple beings, are active, since they carry in them the principle of their changes" (emphasis added).[15] A simple being, for Du Châtelet, is "a being without extension and without particles".[16]

---

[12] G VI 620-621/AG 223.

[13] G VI 598/AG 207.

[14] Jolley (2005, 55–56).

[15] For the observations about Leibniz, see especially Garber (1985; 2009) and McDonough (2013; 2016).

[16] IP §120.

It is unclear from her discussion whether Du Châtelet means to commit herself to the strong claim that simple beings are the only substances, or the weaker claim that simple beings constitute one class of substance (the "real" substances), leaving room for a lesser kind of substance.[17] I remain neutral on this question, and my occasional use of 'simples' to refer to the substances that ultimately underlie or ground phenomena on Du Châtelet's view should not be taken to imply that simples are the only substances for Du Châtelet.

I present two distinct text-based arguments for the claim that Du Châtelet endorses transeunt causation between created substances, one less direct than the other. Let us begin with the less direct argument, which proceeds from a claim about causal interaction between bodies. It is clear that Du Châtelet holds that bodies act upon one another. She writes, for example, in chapter 11, §259 (emphasis added):

> This resistance that all bodies present when one wants to change their current state is the foundation of the Third Law of motion, by which the reaction is always equal to the action.
>
> The establishment of this law was necessary so that bodies might act on one another [*afin que les Corps puffent agir les uns fur les autres*], and that motion, being once produced in the universe, might be communicated [*communiqué*] from one body to another with sufficient reason.

Du Châtelet's claim above implies not only that bodies act on one another, but that a body can "communicate" motion to another body. Du Châtelet is here endorsing a transeunt account of causation that goes beyond a mere counterfactual relation between the states of two bodies: she is endorsing a variety of physical influx.[18]

However, that bodies act on one another, and so stand in causal relations to one another, does not automatically entail that the simple substances that underlie bodies also stand in causal relations to one another. There are at least three ways in which such an entailment might be blocked. First, a body could stand in a causal relation to another body partly or entirely in virtue of something other than the activity of simples. It could, for example, stand in a causal relation to another body partly or wholly in virtue of what God perceives, or in virtue of an a priori condition for the possibility of experience, such as a Kantian category. Second, a body could stand in a causal relation to another body in virtue of causal activity *within* simples (as Leibniz might hold), and not in virtue of any relation that obtains between those simples. Third, even if causal relations between bodies obtain in virtue of relations between simples, the causal relations that obtain between bodies may be explained by *non-causal* relations that hold between simples.

Are there textual or other grounds on which we might rule out the above possibilities? I believe there are. With respect to the first option—on which the causal relations that obtains between bodies obtain in virtue of something other than the activity of simples—Du Châtelet is clear that the sufficient reason for any interaction between bodies lies in simples. She writes:

---

[17] See Carus (2024) for a recent discussion of Du Châtelet's theory of simple beings. Carus argues that Du Châtelet takes a novel, 'faculty-centred' approach to simple beings, in terms of their explanatory role.

[18] Cf. Brading (2018, 85).

> The original reasons for all that happens in bodies lie necessarily in the elements of which they are composed. It follows that the original reason for the connection of bodies to each other, insofar as they coexist and succeed each other, lies in simple beings. So, the connection of the parts of the world depends on the connection of the elements, which is the foundation and the first origin.[19]

This passage suggests that bodies stand in relationships to one another in virtue of the fact that the elements or simples that 'compose' them stand in certain relationships. Barring overdetermination or a disjunctive account, it follows that causal relations between bodies do not obtain in virtue of something other than relationship between simples (such as God's perception or a Kantian category). Du Châtelet's claim that "the original reason for the connection of bodies to each other, insofar as they coexist and succeed each other, lies in simple beings" does not explicitly rule out any other (partial or full) ground for causal relations between bodies. However, the absence of any textual evidence that might point to some other ground suggests—even if it does not entail—that the *complete* reason for the connection of bodies to one another lies in simple beings. The above passage also rules out the second option—one on which the causal relation between bodies obtains in virtue of causal activity within, rather than between, simples—for it states that "the connection of the parts of the world depends on the *connection* of the elements" (emphasis added).

What about the third option? Even if the causal relations between bodies obtain in virtue of relations between simples, it does not yet follow that simples stand in causal relations to one another. That conclusion follows only if it is also *not* the case that a causal relation obtains between bodies in virtue of some other type of relation that obtains between simples. However, there is no textual evidence that suggests that for Du Châtelet, bodies stand in causal relations to one another in virtue of non-causal relations between simples. In the absence of evidence to the contrary, we should at least provisionally conclude that Du Châtelet holds that transeunt causal relations obtain between simples. Thus, transeunt causal interaction between bodies, along with Du Châtelet's premise that "the original reason for the connection of bodies to each other…lies in simple beings" provides an indirect argument for thinking that Du Châtelet endorses transeunt causation between simples.[20]

But we also have a more direct, text-based argument available for the conclusion that Du Châtelet is committed to transeunt intersubstantial causation. Du Châtelet endorses the view that Beings interact with one another. She writes, for example:

---

[19] IP §131. See also IP §133: "So, this connection between our soul and the entire universe comes from the union of the elements among themselves, and from the relationships they all have with one another…".

[20] Others have also noted that Du Châtelet seems to be committed to transeunt intersubstantial causation. Cf. Brading (2018, 63), Stan (2018) and Jacobs (2020, 12–13). Stan writes: "Du Châtelet has no official theory of efficient causation, but we can still infer her most probable view. For one, she claims to follow Wolff on the nature of elements, and he did assert that elements interact by efficient causation…For another, Du Châtelet affirms that *bodies* interact, and that the elements' action and resistance is the sufficient ground of corporeal inter-agency. Short of Leibniz's way—which she did not take—transeunt substantial causation is the only other intelligible ground of outward bodily action" (492).

> And when one considers this Being as placed in the order of things, and linked with the other Beings that surround it, one must show how a Being depends upon its neighbor, and which causes gave actuality to the modes that were simply possible when the Being was considered as isolated and outside the order of things.[21]

Du Châtelet's conception of a Being, however, appears to be more inclusive than that of substance. A 'being' for her is "that which can exist, and whose determinations do not imply any contradiction, whether this Being exists, or whether it is only possible."[22] By contrast, a 'substance' for Du Châtelet is a "[b]eing that endures and is modifiable".[23] A substance is modifiable in virtue of the variation in its modes. By contrast, a being that is not modifiable, and thus not a substance, is an accident, such as a colour. As Du Châtelet writes, "the least modification of this color changes it into another, and it cannot be modified without being changed."[24] Du Châtelet's claim that beings depend upon their neighbors when we consider them as placed in the order of things, and in particular, the claim that merely possible modes had by a being are rendered actual in virtue of the causal activity of its neighbors, strongly suggests that substances, as beings, belong to a causal order.

Moreover, unlike Leibniz, Du Châtelet distinguishes between those features of a being that flow from its essential determinations and those whose actuality depends on an external cause. In the absence of intersubstantial causation, all truths would follow from the "essential determinations" or essential properties of the substance (as a being). But Du Châtelet draws a distinction between the attributes and modes of a being. Attributes, she writes, "have their sufficient reason in the essential determinations"[25] of a being, but these essential properties determine only the *possibility* of modes, since "if the essential determinations contained the reason for the actuality of the modes, the modes would become attributes, since it would be impossible for them not to be found in the Being."[26] For Du Châtelet, the actuality of modes is instead determined through an external cause.[27]

Thus far, I have shown that at the very least, Du Châtelet endorses transeunt causation between the simple substances that ground bodies. Yet, textual evidence also points to a commitment to transeunt causation between souls and other simple substances. In Chapter 7, entitled 'Of the Elements of Matter' Du Châtelet writes:

> For example, I have a very clear idea of this paper, on which I write, and of the pen I use; however, how many dim representations are enclosed and hidden, so to speak, in this clear idea...For the fibers, their differences and their arrangement subsisting, there is no reason why they would not cause impressions on my organs, and consequently representations in my soul; but these impressions being too weak and too compounded, I cannot distinguish them all and they cause dim representations in my soul.[28]

---

21 IP §50.

22 IP §35.

23 IP §52.

24 IP §52.

25 IP §42.

26 IP §43.

27 See, for example, IP §34.

28 IP §132.

The above passage makes it clear that Du Châtelet thinks that objects (such as paper and the fibers that compose them) *cause* impressions on our organs, and *consequently* representations in our souls. The language of causality occurs again in the last line of the quoted passage, where she writes that impressions cause dim representations in our soul.

The claim that representations in our soul are caused by impressions is further bolstered by her claim earlier in §132 that "if at a certain distance the representations that the objects excite in our Soul were to cease, even though the impressions that they make on our senses were to continue, there would be a leap in Nature, which would be contrary to the principle of sufficient reason (§13); for there would be no reason for the clarity of an idea to gradually diminish proportionately to the impression up to a certain point, and then at this point end as if in a leap, even while the reason for its continuation still subsists." Du Châtelet here clearly implies that "the reason" for the continuation of the representation consists in its corresponding impression. If there was no transeunt causation between our organs and souls, and if Du Châtelet instead subscribed to pre-established harmony (even a pre-established harmony restricted to the relationship between the body and the soul), then the sufficient reason for a representation in the soul would surely consist in a prior representation in the soul. Yet, her discussion in §132 makes it clear that she thinks that the representations in the soul can be caused by organs, and the impressions on our organs in turn caused by objects.

In a recent paper Ansgar Lyssy argues that Du Châtelet follows Leibniz in rejecting transeunt causation between simples.[29] However, the textual evidence pointing to this conclusion is not decisive. For example, Lyssy cites §13, where Du Châtelet writes that the "each state in which a being finds itself must have its sufficient reason why this being is in this state rather than in any other, and this reason can only be found in the preceding state." Lyssy takes this passage to suggest that Du Châtelet endorses "primarily self-acting simple beings".[30] However, Du Châtelet's claim occurs in the context of her discussion of the law of continuity according to which "a being does not pass from one state to another without passing through all the different states that one can conceive of between them."[31] As Du Châtelet writes:

> For if there was a state possible between the current state and that which immediately preceded it, the nature of the being would have left the first state without yet being determined by the second to abandon the first. Thus, there would be no sufficient reason why it should pass to this state rather than to any other possible state.[32]

Du Châtelet here derives her law of continuity from the principle of sufficient reason. But her appeal to the principle of sufficient reason (henceforth 'PSR') does not imply that the sufficient reason for a state of a being is to be found *entirely* in the previous state of the being, as opposed to also in an external cause. The passage

---

[29] See Lyssy (2022).

[30] Lyssy (2022, 211).

[31] IP §13.

[32] IP §13.

instead seems to suggest that the contribution of the being to its state *s* is exhausted by the state that immediately precedes *s*. It does not follow from this claim that there is no external cause for the state. In particular, Du Châtelet's claim that the sufficient reason for the state of the being "can only be found in the preceding state" does not, when taken in the context of the law of continuity, imply that there is no external cause that contributes to the state of the being. It instead implies only that insofar as the contribution of the being goes, the sufficient reason for the state is to be found in the state that immediately precedes it. Moreover, "the preceding state" can also be construed broadly so as to include external causes (for example, the state that immediately precedes one where a piece of paper is burning might be one where a candle is held to the piece of paper).[33] It is thus far from clear that Du Châtelet's discussion supports the claim that the sufficient reason for the state of the being consists entirely in the preceding state, considered in isolation from other beings, and it would be too hasty to conclude on the basis of the above passage that Du Châtelet rejects transeunt intersubstantial causation. Additionally, the claim that a created substance actively brings about its own states, even if completely, does not suffice to rule out a commitment to transeunt intersubstantial causation, for the state of a substance may be causally overdetermined.[34] There is no evidence that Du Châtelet explicitly rules out such overdetermination.

Despite rejecting the possibility of transeunt causation between substances, Leibniz (by his own admission) does sometimes talk as though substances act upon one another.[35] For example, in a revealing passage from the *Theodicy* (also discussed by Lyssy), Leibniz writes:

> And in so far as the soul is imperfect and as its perceptions are confused, God has accommodated the soul to the body, in such sort that the soul is swayed by the passions arising out of corporeal representations. This produces the same effect and the same appearance *as if* the one depended immediately upon the other, and by the agency of a physical influence...The same thing must apply to all that we understand by the actions of simple substances one upon another. For each one is *assumed* to act upon the other in proportion to its perfection, although this be only ideally, and in the reasons of things, as God in the beginning ordered one substance to accord with another in proportion to the perfection or imperfection that there is in each.[36]

Could it be that, like Leibniz, Du Châtelet's talk of causation between bodies and between the soul and bodies should not be taken literally? I contend that we lack grounds for treating Du Châtelet's talk of transeunt causation non-literally, for unlike Leibniz, she does not—at least to my knowledge—ever indicate that she intends to talk merely *as if* created substances stand in transeunt causal relations to one another when in fact the causal activity of a simple is restricted to its ability to cause its own

[33] The same point applies to another passage Lyssy discusses (IP §126).

[34] See Amijee (forthcoming) for a discussion of why such overdetermination would have been problematic for Leibniz, given his commitment to the 'collection of requisites' conception of a sufficient reason. By contrast, there is no evidence that Du Châtelet conceived of a sufficient reason for existence in terms of a collection of requisites for an entity's existence.

[35] See also Lyssy (2022, 212).

[36] H 159, §66 (emphasis added).

future states. Such an omission would be glaring and inexplicable, especially since Du Châtelet was likely also aware of Wolff's commitment to transeunt intersubstantial causation (more on this below), and also could not have taken for granted that Leibniz's distinction—that between what is in fact the case and what we suppose to be the case with respect to intersubstantial causation—was common knowledge amongst her readers.

Du Châtelet's position on intersubstantial causation does not appear to be Leibnizian, but it may be Wolffian, at least to some extent.[37] As Jacobs writes, "Wolff…was dissatisfied with Leibniz's substance idealism and moved to a realist view on which fundamental substances are physical elements which lack representational capacities but *are* capable of genuine causal interaction."[38] And as Du Châtelet herself admits, her exposure to Leibniz was mediated through Wolff. She writes in the preface to the *Institutions*:

> I explain to you the principal opinions of M. Leibniz on metaphysics; I have drawn them from the works of the celebrated Wolff, of whom you have heard me speak so much with one of his disciples….[39]

It is also likely that Wolff's knowledge of Leibniz's objections to intersubstantial causation was incomplete. As Watkins (2005, 39) writes, we need to distinguish between the "public" Leibniz and the "private" Leibniz, for the two come apart in significant ways. Many of what we now regard as Leibniz's main texts (his 'Discourse on Metaphysics', 'Primary Truths' and *Monadology*) were not published during his lifetime. And Leibniz writes the following in a July 1714 letter to Nicolas Remond:

> Mr. Wolff has adopted some of my opinions, but since he is very busy with teaching, especially in mathematics, and we have not had much correspondence together on philosophy, he can know very little about my opinions beyond those which I have published.[40]

Thus, it appears likely that Du Châtelet did not set out to consciously reject Leibniz's denial of intersubstantial causation, but was simply influenced by Wolffian ideas more so than Leibnizian ones. The combination of Wolff's limited exposure to Leibniz's considered views and the fact that Du Châtelet's exposure to Leibniz seems to have been mediated through Wolff explains both why Du Châtelet may have had limited (if any) exposure to Leibniz's arguments against intersubstantial causation, and also why she may have been inclined to adopt the Wolffian view on which there is transeunt causation between created substances. Yet, if my earlier arguments succeed, then Du Châtelet's commitment to transeunt intersubstantial causation is more thoroughgoing that Wolff's, for Wolff seems to have endorsed pre-established harmony between substances at least with respect to mind–body relations.[41]

---

[37] Cf. Stan (2018).

[38] Jacobs (2020, 2).

[39] The disciple that Du Châtelet is referring to here is Samuel König, who was a student of Wolff's and Du Châtelet's tutor.

[40] Leibniz (1989b, 657).

[41] See Watkins (2005, 45).

That said, my goal in this paper is not to trace historical influence. Instead, I wish to examine the philosophical motivations for Du Châtelet's commitment to transeunt causation between created substances. In the next section, I will tentatively argue that Du Châtelet's commitment to intersubstantial causation is philosophically undergirded by her commitment to a causal theory of intentionality, along with her distinctive conception of a natural (i.e., non-divine) sufficient reason. On this conception, a natural sufficient reason is such that it can be grasped by us (i.e., finite intelligent beings).[42]

7.3 Tracing Causal Interaction

What philosophical motivation might Du Châtelet have had to endorse intersubstantial causation between simples? I will put forward and defend a hypothesis: Du Châtelet's distinctive account of what qualifies as a sufficient reason, together with her causal theory of intentionality, motivates a commitment to transuent causation between created substances.

Du Châtelet shares with Leibniz, Wolff, and other rationalists a core commitment to the PSR, the principle (broadly speaking) that everything has an explanation. As we will see, however, there is little agreement between Leibniz and Du Châtelet about what the principle requires, beyond this very broad gloss.

Leibniz formulates the PSR in several distinct ways: he sometimes states it as the claim that nothing happens without reason (NE 179, AG 31), or the claim that nothing happens without a cause (AG 31, G. VII. 309), or that every fact or true assertion has a sufficient reason for why it is thus and not otherwise (AG 217), or that nothing happens without a sufficient reason that determines why it is so and not otherwise (AG 210) or that "a reason must be given" for every truth (G. VII. 309), or finally, as the claim that every truth has an a priori proof (G. VII. 309). The above formulations differ in what they take to be the ontological category of the *explanandum*: some take it to be an event while others take it to be a fact or truth. The formulations also differ in what they take to be the ontological category of the *explanans* or sufficient reason: some formulations take it to be a cause, some take it to be a reason, and at least one takes it to be an a priori proof. Finally, they differ in logical form: some formulations are contrastive, whereas others are not.[43]

---

[42] In Amijee (2024) I argue on both textual and philosophical grounds that Du Châtelet is committed to epistemic causal idealism about natural sufficient reasons and causes. As I construe it, epistemic causal idealism about natural sufficient reasons and causes is the claim that it is *constitutive* of any natural sufficient reason and any natural cause that it is knowable by finite minds.

[43] At least one passage from 1686 [as conjectured by Parkinson (1974)] suggests that we should treat them as versions of the same principle. Leibniz writes: "There are two first principles of all reasonings, the principle of contradiction…and the principle that a reason must be given, i.e. that every true proposition, which is not known *per se*, has an *a priori* proof, or that a reason can be given for every truth, or, as is commonly said, that nothing happens without a cause." (G. VII. 309).

What is interesting about these various formulations for my purposes is that none involves reference to what finite intelligent beings or finite minds can grasp. And indeed, as we will see in the next section, Leibniz is explicit that most sufficient reasons (and all sufficient reasons for contingent truths) are beyond *our* grasp. By contrast, Du Châtelet builds reference to a finite mind into her characterization of a sufficient reason.

I make a detailed case for the claim that Du Châtelet's conception of natural sufficient reason and natural cause involves reference to a finite mind in Amijee (2024). But I rehearse some of the highlights of that case for present purposes. First, Du Châtelet often characterizes a sufficient reason as a reason that enables understanding for us. Du Châtelet introduces the PSR in the first chapter of her *Institutions*, where she says in §8 that a sufficient reason is a "reason that makes us understand why this thing is what it is, rather than something completely different". She further writes in §10 that "a cause is good only insofar as it satisfies the principle of sufficient reason" and that a cause that "contains nothing that helps *us to understand* how the vegetation of which I seek the cause operates" (emphasis added) cannot be admitted. On the plausible assumption that a sufficient reason could not enable understanding if we could not access it, it follows that any (natural) sufficient reason is such that it is in principle accessibly by us.[44]

Second, in §8 of the *Institutions*, Du Châtelet argues that the truth of the PSR is needed for the possibility of making judgements on the basis of causal reasoning. Yet the PSR can only play this role if sufficient reasons are, at least in principle if not always in practice, accessible by us. To see why, consider an example Du Châtelet herself provides, where the PSR is employed in causal reasoning. She writes:

> Thus, for example, if I have a ball made out of stone, and a ball of lead, and I am able to put the one in the place of the other in a basin of a pair of scales without the balance changing, I say that the weight of these balls is *identical*, that is the same, and that they are identical in terms of weight. If something could happen without a sufficient reason, I would be unable to state that the weight of the balls is identical, at the very instant when I find that it is identical, since a change could happen in one and not the other for no reason at all…"[45]

Du Châtelet's point in the above passage seems to be that we could not judge that the two balls weigh the same unless the PSR is true. While it is not obvious from the above passage that the cognitive state that Du Châtelet has in mind is knowledge, her claim earlier in the same section that "as soon as one accepts that something may happen without sufficient reason, one cannot be sure of anything"suggests that she has more than mere belief in mind.[46] Moreover, insofar as Du Châtelet intends to argue for the *truth* of the PSR, a charitable reconstruction of her argument depends on

---

This passage is from an essay entitled "*Specimen inventorum de admirandis naturae Generalis arcanis*" ("A specimen of discoveries of marvelous secrets of a general nature").

[44] In Amijee (2024), I argue that natural sufficient reasons and causes, for Du Châtelet, must not only be accessible by us, but *knowable* by us. A defense of this claim goes beyond the scope of the present paper.

[45] IP §8.

[46] IP §8, p. 129.

the claim that the PSR is needed for the possibility of acquiring *knowledge* through causal reasoning, and not mere belief or justified belief. Briefly, this is because the possibility of arriving at a mere belief (or even justified belief) through causal reasoning does not seem to require the *truth* of the PSR, but a mere belief in, or commitment to, the principle.[47]

Now suppose (for *reductio*) that we could not access natural sufficient reasons. We could not then justifiably conclude that the weight of the balls is identical given that we can put one in place of the other without the balance changing, for we could not rule out the obtaining of various defeaters. For example, it may be that the balance is defective and stuck in equilibrium, in which case the sufficient reason for why one ball can be put in place of the other without the balance changing is not that the balls weigh the same, but that the balance is defective. But if this sufficient reason obtained yet could not be accessed by us in principle, we could not justifiably conclude that the two balls weighed the same.[48]

On the assumption that for Du Châtelet natural sufficient reasons are accessible by us, I argue she must also hold that natural *causes* (where a natural cause is just one that is not divine) are accessible by us. This is because Du Châtelet holds that a cause must contain (*contienne*) both the principle of actuality (which brings about the effect) and the sufficient reason (which enables us to grasp how and why the effect occurs).[49] On this view, to grasp a (causal) sufficient reason just is to grasp the cause that functionally 'contains' it.[50] In then follows that if natural sufficient reasons are in principle accessible by us, then so are natural causes.

We are now in a position to ask whether Du Châtelet's commitment to transeunt causation between created substances is in any way informed by her commitment to the claim that natural causes are in principle accessible by finite minds. I will argue that it is. In particular, I will show that Du Châtelet's commitment to a conception of sufficient reason as a reason that is in principle accessible by finite minds, along with her commitment to a causal theory of intentionality, commits her to the possibility of transeunt causation between created substances.

Let us return to Du Châtelet's discussion at IP §132. Du Châtelet there writes that our organs "receive impressions" from "all the motions that happen in the Universe", and in turn cause representations in the soul. It is moreover clear that these representations are *of* the objects that cause impressions on our organs, and thereby

---

[47] See also Amijee (2024) for additional discussion, and in particular the foundational role of the PSR in securing knowledge for Du Châtelet. My argument in this paper, however, does not depend on the claim that, for Du Châtelet, the truth of the PSR is needed to secure the possibility of knowledge acquired through causal reasoning.

[48] My point here is not that Du Châtelet, in introducing her example of the beam balance, is claiming that various defeaters must be ruled out in order for us to be 'sure of anything'. It is rather than the success of her argument for the claim that the truth of the PSR is needed for us to be sure of anything requires that any natural (i.e., non-divine) sufficient reason is such that it is in principle accessible by finite minds.

[49] IP §9.

[50] I defend a functional reading of Du Châtelet's claim that a cause 'contains' a sufficient reason in Amijee (2024).

representations in our souls (see, for example, IP §132, quoted above). Du Châtelet thus expresses a clear commitment to a causal account of intentionality. Our a posteriori access to the external world, for Du Châtelet, therefore depends on what can causally affect us. This now brings us to the question: what *can* causally affect us on her view?

I argued above that natural causes for Du Châtelet are in principle accessible by finite minds, for causes on her view functionally contain sufficient reasons and natural sufficient reasons are, for her, in principle accessible by finite minds. Yet, the class of natural causes is fairly broad, and includes *all* worldly causes, including causes at the level of simple substances. But Du Châtelet, like Leibniz, also holds that what we encounter directly in nature are merely phenomena, and that our access to simple substances is at best confused, where phenomena are "images or appearances that originate from the confusion of several realities."[51] How then might our access to causes at the level of simple substances be possible on Du Châtelet's framework? I propose that a case can be made on Du Châtelet's behalf for our access to causation at the level of substances through a causally mediated inference.

As we saw above, Du Châtelet holds that there is genuine causal interaction between phenomena, and relations at the level of phenomena obtain in virtue of relations between simple substances. Now suppose that a billiard ball b_1 comes into contact with another billiard ball b_2. As I argued in §2, Du Châtelet holds that there is a transeunt causal relation between b_1 and b_2, which obtains in virtue of some causal relation or relations between simples. This grounding relationship between facts about causation between bodies and facts about causation between simples permits us indirect access to causation at the level of simples. Confused perception aside, we do not have access to simples through experience. But we can nevertheless have access to causation at the level of simples by description. For example, we might use the description 'the causal relations that ground the causal relation between b_1 and b_2' to pick out causal relations at the level of simples, and then come to believe (or even know) that such relations obtain by inferring their existence from knowledge that a causal relation obtains between b_1 and b_2. In this way, we might have indirect access to causation between simple substances, even if this access falls short of insight into how precisely simple substances are connected with one another.[52]

Importantly, my claim is not that Du Châtelet endorses the above story for how we might be able to have access to causation between simple substances. My claim is simply that such access is not ruled out on her view, and that we can make sense of how we could have indirect access to the causal activity of simples even if all we perceive in experience are phenomena.

---

[51] IP §154.

[52] Du Châtelet writes at §134: "And there remains to us, of the confused ideas that we have of each of these simple Beings, only an idea of several things coexisting and linked together, without us knowing distinctly how they are linked...". While it may seem that Du Châtelet is here claiming that we cannot have knowledge of or grasp the causal activity of simples, this would be too quick: Du Châtelet is here concerned with our ability to *perceive* simple beings distinctly, and her point cannot be generalized to our access to simple beings more generally.

Divine sufficient reasons do seem to be beyond our grasp in principle, on Du Châtelet's view. Du Châtelet holds that we cannot grasp God's reasons, for the grasp of divine reasons requires an infinite mind. Du Châtelet writes at IP §26:

> It is true that we cannot see all of this grand tableau of the universe, nor show in detail how the perfection of the whole results from the apparent imperfections we believe we see in some parts, for this would require envisaging the entire universe and being able to compare it with all the other possible universes, which is an attribute of the Divinity.[53]

The finitude of our minds—in contrast to the infinitude of God's mind—thus poses an in-principle barrier to our ability to grasp God's reasons (for example, God's reasons for thinking that our world is the best of all possible worlds).[54] By contrast, there is no such in-principle barrier to our ability to grasp reasons at the level of simple substances.

We are now in a position to see how Du Châtelet's commitment to the claim that any natural sufficient reason is such that it is accessible by us, the intimate connection she endorses between causes and sufficient reasons, and her commitment to a causal account of intentionality, all commit her to the possibility (and thus metaphysical coherence) of transeunt causation between substances.

I have argued that Du Châtelet is committed to the in-principle accessibility by us of all natural sufficient reasons, and thus natural causes. It follows that every natural cause is such that our souls (i.e., finite minds) can access, and thus represent it. When supplemented with the causal theory of intentionality, we then get the claim that every natural cause is such that, in virtue of being able to represent it, our souls can stand in a causal relation to it. As mentioned above, simples, for Du Châtelet, are causally active. It then follows that there can be a causal relation between our souls and simples. Thus, Du Châtelet's commitment to the possibility of transeunt causation between created substances, and in particular, between souls and other simples, appears to be philosophically undergirded by her commitments to the causal theory of intentionality and the view on which every natural cause is such that it is accessible by a finite mind.

The above argument can be extended to the possibility of transeunt causation between simples that underlie bodies. Not every simple can *directly* cause representations in a finite mind; most simples must cause representations in a finite mind by

[53] See IP §24 and §26. See also Amijee (2024) for a more detailed discussion of why we cannot access divine reasons.

[54] Carus (forthcoming) argues that Du Châtelet's claim that God created our world because it is the best of all possible worlds is inconsistent with the correct application of her PSR. This is because the PSR, when applied correctly, tells us how and why something occurs, but the mere claim that God created this world because it is the best does not, as a sufficient reason, satisfy this requirement. On Carus's reading, Du Châtelet is inconsistent, and mistakenly endorses the doctrine held by some of her predecessors that our world exists because God created the best of all possible worlds, namely ours. In contrast to Carus's reading, I read Du Châtelet's claims about our ability to grasp sufficient reasons more restrictedly, as applying to only natural (or non-divine) sufficient reasons. On my reading, only natural sufficient reasons must be such that they enable understanding for *us*. Thus, even if an explanation in terms of God's choosing the best, and our world being the best, qualifies as a sufficient reason on Du Châtelet's view yet fails to be illuminating, Du Châtelet's conception of a natural sufficient reason as that which enables understanding for finite minds is not undermined.

causally affecting some intermediate entity, such as our sense organs. As discussed earlier, Du Châtelet holds that objects (such as a piece of paper) impress upon our organs, which then cause representations in our soul. Given that organs are bodies, and on the assumption that every natural cause is such that we can access it in principle, it follows that a natural cause—whether it consists in a body or a simple—can act on the organ. We then have the following possibilities: (1) A body acts on another body (i.e., the organ); (2) A simple acts on a body (i.e., the organ). With respect to the first possibility, we saw earlier that Du Châtelet holds that causal interactions between bodies are ultimately explained in terms of causal relations between simples. But if the causal activity of bodies is ultimately explained in terms of simples, the same is arguably true of simple-body causal relations, and so the simple acts on the organ by acting on the simples that underlie the organ. So, the metaphysical possibility of transeunt causation between non-mental simples follows as long as simples can cause representations in the soul by directly acting upon intermediate bodies (i.e., organs). Thus, Du Châtelet's commitment to her causal theory of intentionality—together with her conception of a natural cause—also requires the metaphysical possibility of transeunt causal relations between non-mental simples, that is, the simples that underlie our organs and other bodies.

I have shown that Du Châtelet's commitment to the metaphysical possibility of transeunt causation between created substances is philosophically undergirded by some of Du Châtelet's other core commitments. First, it is undergirded by her conception of a natural cause, a conception on which every natural cause is such that a finite mind can access it. This conception in turn depends on her conception of a natural sufficient reason, on which every natural sufficient reason is such that a finite mind can access it. Second, it is undergirded by Du Châtelet's commitment to a causal theory of intentionality. In the next and final section, I turn once more to Leibniz.

7.4 Ignorance of Sufficient Reason in Leibniz

Unlike Du Châtelet, Leibniz does not seem to endorse a view on which natural sufficient reasons and causes must be such that finite minds can access them. As we saw above, none of the ways in which Leibniz formulates the PSR involves reference to what a finite mind can access. Indeed, Leibniz is explicit that at least some sufficient reasons *cannot be known* by us. He writes:

> [the principle] by virtue of which we consider that we can find no true or existent fact, no true assertion, without there being a sufficient reason why it is thus and not otherwise, although most of the time these reasons cannot be known to us [*ne puissent point nous être connues*]. (*Monadology* §32; GP 6.612/AG 217)

Leibniz's account of the distinction between the contingent and necessary in terms of an infinitely long or finitely long analysis provides a clue as to why Leibniz might have thought that some sufficient reasons could not be known by us. The account of contingency in terms of an infinitely long analysis that only God can grasp, along with Leibniz's claim (referenced above) that a sufficient reason is an a priori proof (or proof sequence), explains why Leibniz might have thought that sufficient reasons for contingent truths are beyond the grasp of finite minds.[55]

McDonough and Soysal (2016, 2018) argue that Leibniz's account of contingency in terms of an infinitely long analysis does not suggest that contingent truths are just those that lack finite demonstrations. Instead, "Leibniz's core thought is that there is no procedure, no algorithmic process or rule, that we can follow that would allow us to find proofs of contingent propositions."[56] But whether Leibniz's theory of contingency is committed to saying that contingent truths lack finite demonstrations, or whether it instead says merely that a proof for contingent truths cannot be found by appeal to a formal algorithm (i.e., definitions in an ideal language and logic), the theory provides at least one explanation for why Leibniz might have thought that some sufficient reasons—namely, the sufficient reasons for contingent truths—are beyond our grasp. The finitude of our minds would prevent us from completely grasping an infinitely long analysis both because we are not eternal, and because we cannot grasp something with infinite complexity.[57]

One might argue that Du Châtelet's view could be reconciled with Leibniz's if Du Châtelet's commitment to the accessibility of natural sufficient reasons was taken to apply to merely partial, and not complete sufficient reasons. One might then argue that even Leibniz could admit partial grasp of sufficient reasons for contingent truths. Grasp of a few steps of an infinitely long analysis would plausibly constitute such a partial grasp. However, there is no evidence that suggests that Du Châtelet means to restrict the accessibility-by-us of natural sufficient reasons to merely partial sufficient reasons, and textual evidence supports that, for her, non-ultimate explanations qualify as complete sufficient reasons.[58]

Absent any reason—textual or philosophical—to think otherwise, we should not read Du Châtelet as claiming that a non-ultimate cause contains a merely partial sufficient reason, where the complete reason is contained in the ultimate, and in principle inaccessible, cause.

Leibniz appeals to the conceptual independence of substance, and in particular, his conception of a substance as that which corresponds to a complete individual concept—a concept that contains within it all predicates that are true of it[59]—as providing independent logical grounds for rejecting the possibility of intersubstantial causation. He writes:

> We could therefore say in some way and properly speaking, though not in accordance with common usage, that one particular substance never acts upon another particular substance nor is acted upon by it, if we consider that what happens to each is solely a consequence of

---

[55] See, for example, *On Contingency*; AG 29 and *Discourse on Metaphysics* §8, G IV 433/AG 41.

[56] McDonough and Soysal (2016, 454).

[57] Cf. McDonough and Soysal (2016, 452).

[58] See, for example, IP §129.

[59] See, for example, *Discourse* §8 and *Primary Truths*, §32.

> its complete idea or notion alone, since this idea already contains all its predicates or events and expresses the whole universe.[60]

And:

> Strictly speaking, one can say that *no created substance exerts a metaphysical action or influx on any other thing*. For, not to mention the fact that one cannot explain how something can pass from one thing into the substance of another, we have already shown that from the notion of each and every thing follows all of its future states.[61]

Thus, Leibniz's logical conception of a substance as that which is independent of all other created substances seems to underwrite (for Leibniz) his denial of transeunt causation between created substances.

In endorsing transeunt intersubstantial causation, Du Châtelet rejects Leibniz's world apart doctrine. What, then, is Du Châtelet's conception of substance? Du Châtelet presents a positive account of substance in the third chapter of her *Institutions*, and writes at §52:

> …one can define Substance, *that which conserves the essential determinations and the constant attributes, while the modes in it vary and succeed one another*, that is to say, a durable and modifiable subject: for insofar as it has an essence and properties that flow therefrom, endures and continues to be the same, and insofar as its modes vary, it is modifiable.

Prior to presenting her own account of substance, Du Châtelet critiques the accounts of the Scholastics, Descartes and Locke. According to Du Châtelet, the Scholastics defined substance in terms of independence, as "a Being that subsists by itself & is the sustainer of accidents".[62]

What is wrong, according to Du Châtelet, with an account of substance in terms of independence? Du Châtelet writes that "when one wants to know what it is *to subsist by itself, to sustain accidents, and the way in which they are sustained*, one receives nothing in response but new words to define, and to which no distinct idea is attached."[63] Du Châtelet's complaint seems to be that a characterization of substance in terms of independence does not render intelligible what a substance is. If my argument in this paper succeeds, then Du Châtelet's rejection of the independence of substance goes deeper: it is explained ultimately in part by the fact that natural sufficient reasons and natural causes must be graspable by us, a requirement that (given Du Châtelet's other commitments, especially her causal account of intentionality) cannot be satisfied on an independence-based account of substance.

---

[60] *Discourse on Metaphysics* §14, G IV 439-40/AG 47.

[61] *Primary Truths*, A VI, 4,1647/AG 33 (emphasis added).

[62] IP §51.

[63] IP §51.

7.5 Concluding Remarks

I began by showing that Du Châtelet rejects the causal version of Leibniz's world apart doctrine, for she embraces transeunt causation between created substances. In particular, she endorses transeunt intersubstantial causation *both* between souls and the simples that underlie bodies and between the simples that underlie bodies. I then argued that Du Châtelet's commitment to a conception of a natural sufficient reason and cause as that which is in principle accessible by finite minds, along with her causal account of intentionality, can explain her thoroughgoing commitment to the metaphysical possibility of transeunt intersubstantial causation.

By drawing together various threads in Du Châtelet's metaphysical and epistemological framework, I hope to have shown that Du Châtelet's divergence from Leibniz's metaphysics, and potentially Wolff's, is both principled and systematic.[64]

References

Amijee, F. (2024). Du Châtelet's causal idealism. *British Journal for the History of Philosophy*, 1–22.

Amijee, F. (forthcoming). How did Leibniz's God create the World? *Journal of Modern Philosophy*.

Brading, K. (2018). *Émilie Du Châtelet and the foundations of physical science*. Routledge.

Carus, C. (Forthcoming). L'exposition épistémologique innovante du PRS par Du Châtelet - malgré ses incohérences en ce qui concerne la postulation de Dieu et de ce monde en tant que sa création. *Revue de Synthèse*.

Carus, C. (2024). Émilie Du Châtelet's theory of simple beings. *Journal of the History of Women Philosophers and Scientists*, 1–24.

Du Châtelet, É. (1740). *Institutions de Physique*, Paris. [Cited by section number.] Translations by Brading et al. (2018) and Isabelle Bour and Judith Zinsser (2009), available at www.kbrading.org. Cited by section. Abbreviated [**IP**].

Garber, D. (2009). *Leibniz: Body, substance*. Oxford University Press.

Garber, D. (1985). Leibniz and the foundations of physics: The middle years. In K. Okruhlik, & J. R. Brown (Eds.), *The natural philosophy of Leibniz*. Springer.

Harmer, A. (2016). Leibniz's World-apart doctrine. In Y. Chiek & G. Brown (Eds.), *Leibniz on compossibility and possible worlds*. Springer.

Jacobs, C. (2020). Du Châtelet: Idealist about extension, bodies and space. *Studies in History and Philosophy of Science Part A, 82*, 66–74.

Jolley, N. (2005). *Leibniz*. Routledge.

Lascano, M. (2011). Emilie du Châtelet on the existence and nature of god: An examination of her arguments in light of their sources. *British Journal for the History of Philosophy, 19*(4), 741–758.

Leibniz, G. (1923). *Sämtliche Schriften und Briefe, Deutsche Akademie der Wissenschaften*. Multiple volumes in 7 series. Akademie Verlag. Abbreviated [**A**], cited by series, volume, and page.

Leibniz, G. (1978). In C. I. Gerhardt (Ed.) *Die philosophischen Schriften von Gottfried Wilhelm Leibniz* (7 volumes). Georg Olms. Abbreviated [**G**], cited by volume and page.

[64] I am very grateful to Clara Carus and Jeffrey McDonough for their extensive comments on an earlier draft of the paper. I would also like to thank Eric Watkins, Dominic Alford-Duguid, Michael Della Rocca, and Aaron Wells for feedback that improved the paper.

Leibniz, G. (1985). *Theodicy: Essays on the goodness of god, the freedom of man, and the origin of evil* (E. M. Huggard, Trans.). Open Court. Abbreviated [**H**].

Leibniz, G. (1989a). *Philosophical essays* (R. Ariew, & D. Garber, Trans.). Hackett. Abbreviated [**AG**].

Leibniz, G. (1989b). *Gottfried Wilhelm Leibniz: Philosophical papers and letters* (2nd ed.) (L. Loemker, Ed., Trans.). D. Reidel Publishing Company.

Leibniz, G. (1996). *New essays on human understanding* (P. Remnant, & J. Bennett, Eds., Trans.). Cambridge University Press. Abbreviated [**NE**].

Lyssy, A. (2022). Leibnizian Causes in a Newtonian World—Émilie Du Châtelet on Causation. In R. Hagengruber (Ed.), *Époque Émilienne. Philosophy and science in the age of Émilie Du Châtelet (1706–1749)*. Springer.

McDonough, J. (2013). Leibniz's conciliatory account of substance. *Philosophers' Imprint, 13*.

McDonough, J. (2016). Leibniz and the foundations of physics: The later years. *Philosophical Review, 125*(1), 1–34.

McDonough, J., & Soysal, Z. (2018). Leibniz's formal theory of contingency. *History of Philosophy and Logical Analysis, 21*(1), 17–43.

McDonough, J., & Soysal, Z. (2016). Leibniz's formal theory of contingency extended. In U. Beckmann (Ed.), *"Für unser Glück oder das Glück anderer": Vorträge des X. Internationalen Leibniz-Kongresses*. Georg Olms Verlag.

Parkinson, G. (1974). Science and metaphysics in Leibniz's 'Specimen Inventorum'. *Studia Leibnitiana, 6*, 1–27.

Stan, M. (2018). Emilie Du Châtelet's metaphysics of substance. *Journal of the History of Philosophy, 56*(3), 477–496.

Watkins, E. (1995). The development of physical influx in early eighteenth-century Germany. *Review of Metaphysics, 49*(2), 295–339.

Watkins, E. (2005). *Kant and the metaphysics of causality*. Cambridge University Press.

Zinsser, J. (Ed. and Trans.), & Bour, I. (Trans.) (2009). *Emilie Du Châtelet: Selected philosophical and scientific writings*. University of Chicago Press.

Fatema Amijee is Assistant Professor of Philosophy at the University of British Columbia. A primary focus of her work is the Principle of Sufficient Reason.

Chapter 8
Infinity for Us: Du Châtelet on the Labyrinth of the Continuum

Gabriel Müller

8.1 Introduction

> There are two famous labyrinths, in which our reason often gets lost: One regards the great question of the free and the necessary, in particular in the production and the origin of evil; the other consists in the discussion of continuity and of the indivisibles which seem to be its elements, a discussion which must involve the consideration of the infinite. The first labyrinth embarrasses all of humanity, the other bothers only the philosophers.[1]

The two philosophical problems sketched by Leibniz in this passage occupied him throughout his career. He called them "labyrinths," here in the *Theodicy* as well as in many earlier texts, to express that they were confusing and difficult to resolve, but also of central importance. Here I am interested mainly in the second labyrinth, which, as Leibniz alludes in the quotation above, consists in the attempt of formulating a coherent account of continuous quantities. It arises from the combination of two intuitions: On the one hand, for any extended entity to be continuous means that it has no determinate parts, therefore it would seem to be infinitely divisible. On the other hand, it also seems plausible that material wholes are constituted by their parts, which is however only possible if the parts in question are determinate, i.e., not themselves continuous. It therefore seems impossible to decide whether extended entities have determinate parts or not.

[1] Leibniz (1720, x): "Il y a deux Labyrinthes fameux, où notre Raison s'égare bien souvent: l'un regarde la grande Question du Libre & du Nécessaire, sur tout dans la production & dans l'Origine du Mal; l'autre consiste dans la discussion de la Continuité, & des Indivisibles, qui en paroissent les Elemens, où doit entrer la consideration de l'Infini. Le premier embarasse Presque tout le Genre humain, l'autre n'exerce que les Philosophes."

G. Müller (✉)
University Library Basel, Basel, Switzerland
e-mail: gabriel.mueller@unibas.ch

C. Carus and J. K. McDonough (eds.), *Émilie Du Châtelet in Relation to Leibniz and Wolff*, Women in the History of Philosophy and Sciences 23,
https://doi.org/10.1007/978-3-031-80541-7_8

This puzzle occupied Leibniz until the end of his life, and it has occupied many scholars of Leibniz over the years.[2] It also had a more immediate impact in the first half of the eighteenth century, however, namely on the debates about Leibnizian philosophy generally, and more specifically about the possibility of using Leibniz' system as a foundation for physics. It is a part of this reception that I will investigate in this paper. The focus will be on Du Châtelet's view on the nature of the labyrinth and on the way out of it, but that view will become more clear trough a comparison with those of two of Du Châtelet's contemporaries, namely Christian Wolff and Leonhard Euler.

Wolff's influence on the metaphysics of the *Institutions* is well documented: According to Du Châtelet herself, the writings of the "celebrated Wolff" (Du Châtelet, 1740, avant-propos, 13) were her main source of Leibnizian philosophy in the wide sense. Recent work by Marius Stan has shown, moreover, that her metaphysical position is closer to that of Wolff than that of the late Leibniz on a number of important points (Stan, 2018). The rest of this paper will confirm this for the issues of the continuum in particular, though it will be seen that some of Leibniz' published work could plausibly have influenced her position as well.

In addition to reconstructing and comparing the views of Du Châtelet and Wolff on the continuum, I will also confront them with the arguments of Euler, who is opposed to them in almost every respect. Following the winding path of the labyrinth's reception from Leibniz to Wolff to Du Châtelet to Euler is interesting in its own right, but the contrast with Euler will also be helpful for the reconstruction of Du Châtelet's view, because it reveals that certain aspects of the conceptions of Du Châtelet and Wolff are designed to respond to arguments like Euler's.

Euler is best known as a mathematician, but he also wrote a popular exposition of Newtonian natural philosophy in the form of his *Letters to a German Princess*.[3] Euler discusses motion theory, gravity, magnetism and electricity, acoustics and especially optics, but also finds time to address more general philosophical topics, among them the relation of body and soul and the basics of epistemology and logic. Euler's *Letters* therefore cover a broader range of scientific topics than Du Châtelet's *Institutions*, whereas she spends more time on the philosophical topics in the modern sense. Notwithstanding the differences in emphasis and method between the *Letters* and the *Institutions*, the two books transport similar visions of the scope of natural philosophy. As has been noted previously, the scientific (as opposed to the more foundational philosophical) material covered in the *Institutions* is basically that of Newton's *Principia mathematica*. This is far from being the norm, since other introductions to natural philosophy published before 1740 take the discipline to encompass a much broader spectrum of phenomena. This is particularly true of books published in France, like the influential *Traité de physique* of Jacques Rohault, to

[2] Notably, Kant in the Second Antinomy, A 434-441/B 462-469. Some modern scholars that have engaged with the problems of the continuum in Leibniz are Wilson (1989, 74–77), Levey (1998), Arthur (1998, 2018), Phemister (2005, 105–132), De Risi (2007), Garber (2009, 155–166).

[3] I quote the English translation of the *Letters* in (Euler, 1802); the original French is (Euler, 1768). On Euler and the *Letters*, see (Calinger, 1976, especially 212–13 and 225–27).

which Du Châtelet also compares her own project in the preface of the *Institutions* and which includes sections on human anatomy and meteorology. When compared to this Cartesian model of a physics textbook, there are clear parallels between Euler's *Letters* and Du Châtelet's *Institutions*, despite the fact that Euler includes some topics that Du Châtelet ignores.[4] The clearest parallel between them is their intended audience: Both works are introductions to natural philosophy written in French, directed at educated laypeople who have some knowledge of philosophy but next to none of advanced mathematics.

Euler's project is therefore comparable to Du Châtelet's, in the sense that they have a shared vision of what physics or natural philosophy encompasses. They differ strongly, however, when it comes to the philosophical framework in which they situate their physics. Especially in those sections in which he writes about the philosophical foundations of matter theory, Euler's main opponents are broadly Leibnizian philosophers whom he calls "monadists." The identity of these opponents remains somewhat vague: Though Wolff is mentioned once (in vol. I, letter 76), Euler most often writes in the plural and leaves open the possibility of internal differences in the monadist camp. Du Châtelet, in contrast to Wolff, is not mentioned by name, but there is reason to believe that she is at least one of the intended targets of Euler's critique of the monadists: In addition to the aforementioned parallels between the aim and style of their works, there is a letter from Euler to Du Châtelet which documents that he had received a copy of the *Institutions* from her. In this letter, of which unfortunately only an undated draft is preserved, Euler commends the marquise on her book and on her treatment of hypotheses in particular (1963, 278–279). As we shall see, moreover, Euler's descriptions of the monadist positions can often be read as reasonably accurate (though uncharitable) descriptions of positions found in the *Institutions*.

8.2 Distinguishing Geometrical and Physical Extension

Euler's stance on the structure of the continuum is that there is no need for extended bodies to be constituted by indivisible parts of any kind, neither spiritual nor material. According to him, anything that is extended can always be further divided, without thereby presupposing any ultimate particles (Euler, 1802, II, 41). He denies, in other words, that there needs to be any determinate and discrete entities in order for there to be a continuum. There is one argument against postulating indivisibles in particular that he repeats multiple times, namely that it would require one to draw an ultimately incoherent distinction between geometrical and physical divisibility. He suggests that

[4] See (Janiak, 2018, 61) for the comparison between Du Châtelet, Newton, and other contemporary works on natural philosophy. Suisky (2012, 117) has also called the *Institutions* a "forerunner" of the *Letters*. Rohault's *System* is mentioned in (Du Châtelet, 1740, avant-propos, 3). As Hutton (2012, 84) points out, Du Châtelet owned Rohault's book in the version annotated by Samuel Clarke, either in the original English or in a Latin translation.

there are certain contemporaries of his who make such a distinction (Euler, 1802, II, 25–26):

> There are however philosophers, particularly among our contemporaries, who boldly deny, that the properties applicable to extension, in general, that is, according as we consider them in geometry, take place in bodies really existing. They allege that geometrical extension is an abstract being, from the properties of which it is impossible to draw any conclusion, with respect to real objects.

But, Euler argues (1802, II, 33), "if divisibility in infinitum is a property of extension in general, it must of necessity likewise belong to all individual extended beings." Even if geometrical extension is an abstract and general notion, our conclusions about it must also apply to the extended physical bodies in which it is instantiated. Euler shows that by means of simple geometrical operations, any given line, surface or body can in principle be divided; the same must therefore be true of the physical bodies that instantiate these abstracta. He concludes that if there are obstacles preventing us from dividing bodies beyond a certain limit, they must be of a merely practical nature, which is why "[i]n speaking of the divisibility of body, we must carefully distinguish what is in our power, from what is possible in itself" (39).

Chapters 7–10 of Du Châtelet's *Institutions* contain an extended discussion of how the physical world and visible bodies arise from simple substances. This Chapter is dedicated to an investigation of the divisibility of matter specifically. It is there that Du Châtelet first mentions the labyrinth of the continuum, and her solution would seem to consist in drawing precisely the kind of distinction that is so condemned by Euler. This is how she presents the problem:

> The majority of philosophers, having confounded the abstraction of our mind with the physical body, wanted to prove the infinite divisibility of matter through the arguments of the geometers about the divisibility of lines that is taken to infinity, which has given occasion to that famous labyrinth of the divisibility of the continuum which has so embarrassed the philosophers. But they would have been spared all difficulties which that divisibility entails if they had taken care never to apply arguments about the divisibility of the geometrical body to natural and physical bodies.[5]

This passage leaves open what exactly the labyrinth of the continuum consists in, though it seems clear that in the phrase itself Du Châtelet is alluding to Leibniz'

---

[5] Du Châtelet (1740, ch. 9, 181): "La plûpart des Philosophes ayant confondu les abstraction de notre esprit avec le Corps Physique, ont voulu démontrer la divisibilité de la Matiére à l'infini, par les raisonnemens des Géometres sur la divisibilité des lignes qu'on pousse jusqu'à l'infini; ce qui a donné lieu à ce labyrinthe fameux de la divisibilité du continu, qui a tant embarrassé les Philosophes: mais ils se seroient épargné toutes les difficultés que cette divisibilité entraîne, s'ils avoient pris soin de ne jamais appliquer les raisonnemens que l'on fait sur la divisibilité du Corps Géometrique, aux Corps naturels & Physiques." The explicit reference to the labyrinth occurs only in the original 1740 version of the paragraph and is omitted from the corresponding passage in the 1742 edition. See Sect. 8.4 below on the impact of the changes between the two editions. Note also that in both editions, a passage in §87 mentions the "labyrinths of errors" in which those that fail to distinguish between the mathematical and the physical perspective get lost, there in the context of the ideality of space.

formulation in the *Theodicy*-passage quoted above. We do learn that the philosophical problem arises from misguided attempts to prove that matter is infinitely divisible. To solve this problem, Du Châtelet claims, one simply needs to refrain from applying geometrical arguments to bodies. Her reasoning for why this is possible is not contained in the passage from §168 just quoted. This is of course a central question, and I will come back to this question in the second half of the paper. Before that, however, it will be helpful to reflect on the history of the distinction between geometrical and physical divisibility. This will show, I hope, that parts of Euler's remarks impute a view of this distinction to the monadists that is quite misleading. Not only is it held neither by Du Châtelet nor by Wolff nor by Leibniz, but it would have been entirely discredited in the eyes of anyone familiar with the basics of seventeenth-century matter theory.

8.3 Corpuscularianism

Metaphysically indivisible, extended atoms were a fringe position at the beginning of the early modern period, and they remained so until the eighteenth century and beyond. A case in point is Descartes: When he introduced his new natural philosophy, he was adamant that while it was a refutation of Aristotelianism, it was by no means a revival of ancient atomism either. On the contrary, he combined his defense of corpuscles with a new line of argument against physical atoms: By means of his identification of physical substance and *res extensa*, Descartes was able to infer the non-existence of the vacuum and the infinite divisibility of physical bodies.[6]

In the generation after Descartes, however, many remained doubtful whether the identification of physical substance and extended thing could ever provide a sufficient foundation for natural philosophy. True Cartesians defended it, of course, an influential example being the aforementioned Rohault (1723, I, ch. 7). One alternative was championed by Robert Boyle, who emphasized the simplicity and power of explanations using corpuscles over the need for a metaphysical proof of their existence, e.g. in his *Of the Excellency and Grounds of the Corpuscular or Mechanical Philosophy* (1772, IV, 68–69).[7] Others, of which Leibniz is the most obvious example, argued that a natural philosophy that deserved its name needed to be founded on metaphysical principles. Leibniz's position on the exact nature of these metaphysical principles evolved over the course of his life, but at least in the later parts of his career he held them to be monads, i.e., unextended, incorporeal entities endowed with perceptions and appetites.

It is this Leibnizian project that Du Châtelet is engaged in, as she herself explains by means of an excursus on the history of matter theory. At the beginning of her Chap. 7 on the "Elements of matter," she gives a brief history of doctrines on the

---

6 *Principia philosophiae*, II, 20, in (Descartes, 1897, VIIIa, 51–52). See (Garber, 1992, ch. 5, 117–34).

7 Cf. also (Boyle, 1772, I, 308–11) on the relative value of a priori versus empirical arguments.

smallest elements of material bodies, in order to explain the need for unextended simple substances. In her telling, there were four relevant stages to this development. In ancient times, Aristotle synthesized the thoughts of his predecessors into the system of the four elements, each of them simple in the sense of not being composed of anything of a different nature than itself. Descartes then replaced Aristotle's four elements with his three types of corpuscles, these corpuscles being differentiated by their size and shape. Thirdly, post-cartesian natural philosophers dropped the differentiation into three separate types of particle and simply assumed that matter is "a uniform and similar mass, without any internal difference, but the small parts of which have such different shapes and sizes that the infinite variety that reigns in this universe can result from them."[8] The fourth phase in the development of matter theory begins with Leibniz' insight that "these atoms did not give him the reason for the existence of the extension of matter"[9] and that extension itself needed to be founded in unextended monads. The doctrine of the elements of matter, as Du Châtelet proceeds to lay it out in the rest of Chap. 7, is presented in §119 as a summary of the ideas of both Leibniz himself and Wolff regarding this Leibnizian problem.[10] Like Leibniz, therefore, Du Châtelet is convinced that metaphysical certainty about the ultimate structure of extended bodies is required for a complete natural philosophy. Also like Leibniz, she proposes unextended substances as the best means to achieve such certainty, though her simple substances differ from monads, notably by their lack of perception.

With regards to the continuum, there is another crucial difference between their views. In the case of Leibniz, the problem that he calls the labyrinth of the continuum results from the combination of his metaphysical views with his views on the divisibility of bodies.[11] Rather than positing any absolutely indivisible atoms or even merely physically indivisible corpuscles, Leibniz holds that all bodies are actually divided to infinity. His reasoning is most explicit in an unpublished fragment:

> Created things are actually infinite. For any body is actually divided into multiple parts, since any body is acted upon by other bodies. And any part of a body is itself a body, [which follows] from the very definition of body. And so bodies are actually infinite [in number] or there are more bodies than there are unities in any given number. The consequence is evident, since if an arbitrary division was given into only two parts, neglecting all others, and if of these two parts only the second one was given as subdivided, the subdivision of the second one would produce as many parts as there are divisions, as in $A = L + B$, $B = M + C$, $C = N + D$. It is evident that three divisions produce at least three distinct [bodies] L, M, and N.[12]

[8] Du Châtelet (1740, 7, 130): "[...] une masse uniforme & similaire, sans aucune différence interne; mais dont le petites parties ont des formes & des grandeurs si diversifiées, que la variété infinie qui règne dans cet Univers peut en résulter."

[9] Du Châtelet (1740, 7, 131): "[...] ces Atomes ne lui donnoient point la raison de l'étendue de la Matière."

[10] Ibid.

[11] For the following two paragraphs, I draw on (Arthur, 2018), especially 42–44.

[12] Leibniz (1970, VI, 4, 1392–93): "Actu infinitae sunt creaturae. Nam quodlibet corpus actu subdivisum est in plures partes quia quodlibet corpus ab aliis corporibus patitur. Et quaelibet pars

The argument is the following: Let us assume that we have the smallest actual division, that is, a finite number of extended bodies that are not actually divided any further. In this situation, each of these smallest bodies would be acted on by other bodies. According to one of the basic principles of Cartesian physics, any physical interaction results in the relative motion of two parts of matter, which are thereby individuated into separate bodies, and to be actually divided is nothing else except to be acted on by multiple other bodies. In the Cartesian version of a mechanical universe, furthermore, where all of space is filled with bodies and causation consists in the pushing of particles against particles, the motion of any one body indirectly affects all others. Each of the minimal particles that exist on the supposition, therefore, is under the effect of a multitude of motions, each one originating in one of the other minimal particles. Each of these motions, in turn, divides the supposedly minimal particle into smaller bodies. Finally, as Leibniz shows, an infinity of ever-smaller divisions will result in an infinite number of parts, because every new division produces at least one new part.

Though this argument relies on two distinctly Cartesian principles (i.e., universal interaction and individuation of bodies through relative motion), Leibniz takes it to a more radical conclusion than Descartes. As multiple scholars have pointed out, an important inspiration for Leibniz' transformation of Descartes' view of physical division was *Principles of Philosophy* II, 34 (Descartes, 1897, VIIIa, 59–60). In that section, Descartes had used the circular motion of a fluid through a pipe of varying width to argue for infinite divisibility. Leibniz thought the argument was admirable, but that Descartes had not fully appreciated that it showed the infinite actual division of matter, not merely its infinite divisibility.[13]

Leibniz' full account of the composite and its divisions was not widely known among his contemporaries, since he expressed most of it in private notes or letters. The claim that bodies are actually infinitely divided, however, certainly was. Leibniz also upholds it in his publicly known works, for example in the *Monadology*:

> Each portion of matter is not only divisible to infinity, as the ancients recognized, but even actually subdivided without end, each part again into parts, each one of which has its motion.[14]

Leibniz therefore defends not only infinite divisibility, but infinite actual division of bodies, and he does so quite publicly. This is not the case for either Wolff or Du Châtelet, as we shall see in the next section.

---

corporis est corpus ex ipsa corporis definitione. Itaque actu infinita sunt corpora seu plura reperiuntur corpora, quam sunt unitates in numero quocunque dato. Consequentia patet; quia si quaelibet divisio poneretur facta in duas tantum partes, neglectis reliquis; et ex his duabus partibus altera tantum subdivisa poneretur, subdivisione alterius tot minimum prodirent partes, quot fiunt divisiones, ut $A = L + B$, $B = M + C$, $C = N + D$, patet ex tribus divisionibus prodire minimum tria diversa, L, M, N."

[13] See (Levey, 1998, 53) and (Arthur, 2018, 42) for the reception of *Principles* II, 34 in Leibniz.

[14] Monadology §65, in (Leibniz, 1875, VI, 618): "chaque portion de la matiere n'est pas seulement divisible à l'infini, comme les anciens ont reconnu, mais encor sous-divisée actuellement sans fin, chaque partie en parties, dont chacune a quelque mouvement propre." A similar statement is found in the *Theodicy* (Leibniz, 1720, §195).

8.4 Material Atoms and Atoms of Nature

Wolff's doctrine on the divisibility of matter, in its most elaborate form, can be found in his Latin treatise on natural philosophy, the *Cosmologia generalis* (Wolff, 1737).[15] According to Wolff, there exist what he calls "material atoms," extended particles that are not divisible by natural causes. From these material atoms, Wolff distinguishes the absolutely indivisible "atoms of nature":

> An atom of nature is what is indivisible in itself, and so lacks any parts into which it could be divided. A material atom, however, is that which is divisible in itself, but for the division of which no causes that exist in nature are sufficient.[16]

Though material atoms are not divisible by natural means, they *are* divisible in the geometrical perspective. As Wolff's phrasing suggests, they could be divided by God's omnipotence. Wolff also calls the material atoms "primitive corpuscles" because they are the basic building blocks of material things.[17] In contrast, simple substances are absolutely indivisible atoms of nature, because they have no extension. It is this distinction between primitive corpuscles or material atoms on the one hand and simple substances or atoms of nature on the other that, according to Wolff, provides the way out of the labyrinth of the continuum:

> Commonly people acknowledge nothing but derivative corpuscles. For this reason, they consider every corpuscle as if it is composed out of smaller ones, and therefore do not acknowledge any mode of composition except through other explicable corpuscles. But then they fall into the labyrinth of composition and division, from which there is no way out, owing to the fact that one never reaches a last term in division, nor a last reason in composition.[18]

To Euler's accusations that the monadists deny that real bodies are infinitely divisible, therefore, Wolff would have been able to concede that all extended bodies are divisible—provided that the dividing power is supernatural. God is able to divide the material atoms into their constituent simple substances. If the dividing powers are limited to natural forces, on the other hand, the primitive corpuscles are indivisible. The simple substances are not divisible even by God, but since they are without extension the geometrical argument does not apply to them.

---

[15] This is the second edition of Wolff's work. It was first published in 1731.

[16] Wolff (1737, §168): "Atomus Naturae dicitur, quod in se indivisibile est, adeoque partibus destituitur, in quas resolvi possit. Atomus autem materialis appellatur, quod in se divisibile, sed cui dividendo non sufficiunt aliquae causae in rerum natura existentes".

[17] Wolff (1737, §229): "Primitive particles are those for which no reason can be assigned of [their] composition except to the elements."—"Corpuscula primitiva sunt, in quibus nulla compositionis ratio assignari potest, praeterquam in elementis".

[18] Wolff (1737, note to §229): "Vulgo non agnoscunt nisi corpuscula derivativa: unde corpusculum omne tamquam ex aliis minoribus compositum considerant, neque adeo alium compositionis modum agnoscunt, nisi qui per alia corpuscula explicabilia. Sed tum in labyrinthum compositionis ac divisionis continui incident, ex quo exgredi non datur, propterea quod nunquam in divisione ad terminum ultimum, nec in compositione ad rationem ultimam perveniatur." Translation from (Arthur, 2018, 29).

Does Du Châtelet have an analog to Wolff's material atoms? I would argue that she does, though she expresses her position more clearly in the second edition of the *Institutions* than in the first one. In the original 1740 edition of the *Institutions*, the account of divisibility is contained in a single long paragraph, number 172. There, Du Châtelet argues that even though the notion of absolutely indivisible extended particles is contradictory, the existence of naturally indivisible particles is very likely:

> We have seen above that atoms or uncuttable parts of matter cannot be admitted as long as we regard them as simple, primitive and not reducible, because it is not possible to give any sufficient reason for their existence. But as soon as we recognize that they have their origin in the simple beings, it is very well possible to admit them, for it is highly possible, and experience renders it highly believable, that there are in the universe a certain determined number of parts of matter that nature never resolves into their principle.[19]

Du Châtelet holds, therefore, that some kind of extended particles exist, and that they can only be postulated without contradiction when it is assumed that they are founded on the simple beings. Furthermore, the number of these particles in the universe is finite. Otherwise, however, her explanation does not say much about the nature of the corpuscles. One open question is how exactly the foundation of the particles on the simple beings works; another question is why it should be true that the particles are not dissolved in the course of nature. Du Châtelet answers the second question in §173, where she refers to the Leibnizian doctrine that "conspiring motions are the cause of cohesion."[20] But the question of divisibility becomes even more urgent when, in §§174–55, she emphasizes how prevalent continuity is in nature. §175 in particular takes this so far that it becomes difficult to see what place might remain for extended indivisibles. Du Châtelet writes the following:

> These considerations show us that the subtlety of the parts of matter is inexpressible, and that there is nobody who can ever determine the number of parts of which a grain of sand is composed, since that number surpasses our imagination and everything that we can invent; and since reason shows us that this division has no limits, and that matter does not cease to be divisible insofar as it is matter, we may say that in relation to us, matter is not only divisible, but divided to infinity, even though its divisions really have limits; for these limits are so remote that they extend to infinity for us, because infinity for us is a quantity that no number can express.[21]

---

[19] Du Châtelet (1740, 185): "On a vû ci dessus que les atômes ou parties insecables de la Matiére sont inadmissibles, quand on les regarde comme des Matiéres simples, irrésolubles & primitives, parce qu'on ne peut point donner alors de raison suffisante de leur éxistence; mais lorsqu'on reconnoît qu'ils tirent leur origine des Etres simple, on peut fort bien les admettre: car il est très-possible, & les expériences rendent très-vraisemblable qu'il y a dans l'Univers un certain nombre déterminé de parties de Matiére, que la Nature ne résout jamais dans leur principe, qui restent indivisées dans la constitution présente de cet Univers, & que tous les Corps qui le composent, resultent de la composition & de la mixtion de ces particules solides; ensorte qu'on peut les regarder comme des Elemens doués de figures, & de différences internes, qui resultent de leurs parties".

[20] Du Châtelet (1740, 188): "… les mouvemens conspirans, sont la cause de la cohésion, selon M. Leibnits."

[21] Du Châtelet (1740, 191): "Ces considerations nous montrent que la subtilité des parties de la Matiére est inexprimable, & qu'il n'y a personne, qui puisse jamais déterminer le nombre des parties, dont un grain de sable est composé, puisque ce nombre passe notre imagination, & tout ce

The view that is presented here seems rather like the one favored by Euler: Pure reason shows us that as long as there is matter, it always remains possible in principle to divide it, and though any real division has its limits, these are merely practical. But how can such a view be reconciled with what we know about the rest of Du Châtelet's position? How can matter be infinitely divided, even if it is only "in relation to us," if the number of simple elements is finite? The solution to this interpretative puzzle, in my view, is that even the "divisibilist" talk in the passage just discussed is not meant to imply actual infinite division. Evidence for this comes from the changes she made to this part of the *Institutions* for the second edition of the book, during which she completely rewrote the paragraphs from §172 to the end of Chap. 9. In the course of these revisions, she replaced the misleading formulations, re-structured her argument slightly, and explained more clearly the structure of the extended particles. The relation of her extended indivisibles to Wolff's material atoms also becomes more explicit.

We have seen that in the first edition, the labyrinth of the continuum is mentioned in §168. For the second edition, the sentence in question is cut from §168, and instead the labyrinth is mentioned at the beginning of the revised section, at §172. The original version of that paragraph had consisted of a number of plausibility arguments for the existence of stable particles. In the revised version of the same paragraph, Du Châtelet instead emphasizes that we never encounter direct evidence of uncuttable particles: "However far we push our divisions, there still remains extension."[22] She now argues that it is this physical experience, together with the misguided application of geometry to bodies, that leads philosophers into the "labyrinth of the composition and infinite division of the continuum."[23]

In the revised Chap. 9, she focuses on the extended particles: Almost directly after the introduction of the labyrinth, she gives a paraphrase of Wolff's distinction of primary and secondary particles. What she emphasizes most of all in the subsequent discussion is how incredibly small and numerous even the derivative particles are. Among the experiments cited is the fact that a small amount of gold can be hammered into a very large leaf. The point of these illustrations is to show that even though according to her theory, matter is constituted by discrete, extended entities, it is no contradiction that it might *seem* like a continuous quantity in all aspects that are accessible to human experimentation.

The text of the second edition makes quite clear that Du Châtelet embraces the same hierarchy of particles as Wolff does: Simple elements at the very bottom (unextended and absolutely indivisible), the extended primary particles formed out of the

---

que nous pouvons nous figurer; & comme la raison nous montre que cette division n'a point de bornes, & que la Matiére ne cesse point d'être divisible tant qu'elle est Matiére, on peut dire que par rapport à nous la Matiére est non seulement divisible, mais divisée à l'infini, quoique réellement ses divisions ayent des bornes; car ces bornes sont si reculées, qu'elles s'étendent jusqu'à l'infini pour nous, car l'infini pour nous est une quantité qu'aucun nombre ne peut exprimer."

[22] Du Châtelet (1742, 194): "que quelque loin nous poussions nos divisions, il reste toujours de l'étendue."

[23] Du Châtelet (1742, 195): "ce labirinte de la composition, & de la division infinie du continu [...] dont les etres simples peuvent seuls nous tirer."

simple elements (extended, naturally indivisible), then many tiers of secondary particles (divisible in principle, though most of them too small to be divided in practice). In light of these statements, the "infinite division" mentioned in the original version of §175 must be a figure of speech: When Du Châtelet writes that matter is divided to infinity in relation to us, she means it in the sense that for any division of matter that human artifice can achieve, the primary particles are smaller still.

As we have seen, Euler reminds the monadists that "[i]n speaking of the divisibility of body, we must carefully distinguish what is in our power, from what is possible in itself" (Euler, 1802, II, 39).[24] It should be clear by now that none of our monadist authors are in any need of such a reminder. Leibniz certainly was not, and in any case he was almost exclusively concerned with metaphysical indivisibility. The two types of indivisibles introduced by Wolff and Du Châtelet, for their part, are based on precisely the distinction Euler accuses them of forgetting.

8.5 Extension as Real Phenomenon

The core of Euler's critique, however, would seem to be made stronger by the fact that Wolff and Du Châtelet postulate two kinds of indivisibles. His main point, as we have seen, was that the monadists were forced to claim that the properties of geometrical extension, and infinite divisibility in particular, do not "take place in bodies really existing" (1802, II, 26). And indeed, according to both Wolff and Du Châtelet, extended bodies are only infinitely divisible "for us," that is to say, as far as natural and human powers are concerned. The theoretical limit of natural division are the primitive corpuscles, and though these are still divisible by divine means, each of them only contains a finite number of simple substances and is therefore not *infinitely* divisible. It is therefore legitimate to ask, as Euler does, how the infinite divisibility of geometrical extension fits into the picture. In letter VIII, he argues that since geometry shows that extended atoms are impossible, the only indivisible particles that can be coherently conceived must be without extension (Euler, 1802, II, 29):

> Whoever is disposed to deny this property of extension [i.e., its infinite divisibility], is under the necessity of maintaining, that it is possible to arrive at last at parts so minute as to be unsusceptible of any farther division, because they ceased to have any extension.

According to Euler, then, there are only two alternatives: Either bodies are infinitely divisible, without simple substances constituting them, or there is a limit to division, namely the unextended simple substances. The former position is of course Euler's own, while the latter is that of the monadists. In his view, it implies a counter-intuitive picture of the division of bodies (1802, II, 35):

> [The monadists] insist, that this division extends only to a certain point, and that you may come at length to particles so minute, that, having no magnitude, they are no longer divisible.

---

[24] See above, Sect. 8.2.

> These ultimate particles, which enter into the composition of bodies, they denominate simple beings, and monads.

According to Euler's characterization of the monadist position, therefore, an extended thing can be divided into simple beings by means of a finite number of divisions, which amounts to the claim that a finite number of unextended parts can constitute an extended body. Against this, Euler argues along the following lines: Assume that after a body with a length of one inch is divided into one thousand parts, one arrives at indivisibles; since these are necessarily unextended points, the length of the original body is a thousand times zero, which is still zero (1802, II, 29–30). Essentially the same argument is repeated in a later letter (1802, II, 42). What this argument shows is that if the indivisible simple beings are to constitute an extended body, they cannot do so in the way in which parts constitute a whole.

What the monadists therefore need to do in response is to deny that the relation of extended bodies and unextended indivisibles is a relation of parts and wholes, and to give an alternative explanation. This is precisely what they do, since they all hold that extension is a phenomenon founded in the simple beings. Much has been written on the exact meaning of this doctrine in Leibniz, and especially on the question of the development of Leibniz' view over time, the opinions of scholars diverge. For our present purposes, it is enough to know that the monads of the late Leibniz were "not parts, but the foundations of the phenomena," as he writes to De Volder in June 1704 (Leibniz, 1976, 536). Du Châtelet echoes this Leibnizian language when she writes in §133 of the *Institutions* that "this union [of the elements] is the foundation of the union of the parts of matter," and in §134 that extension "is but a phenomenon, an abstraction from multiple real things."[25] In the continuation of that paragraph, she uses a counterfactual to explain what this means for the reality of extension:

> And so, if we were able to see distinctly all that composes extension, that appearance of extension that seems so obvious to us would disappear; and our soul would have apperceptions only of simple beings existing outside one another; just as if we were to distinguish all the little portions of matter moving in different ways that compose a portrait, that portrait, which is a phenomenon, would disappear for us.[26]

In other words, extended bodies are real insofar as they can be identified with an assemblage of simple beings, but they are also phenomena, because some of their properties depend on certain features of human perception and imagination. If our minds and eyes were less prone to introduce such distortions, that is, if we were able to see the world as it really is, we would only need to refer to the simple beings themselves. The impression that the extended bodies have any existence that is not simply the existence of the simple beings would therefore disappear.

---

[25] Du Châtelet (1742, 154): "cette union est le fondement de celle des parties de la matière." (1740, 149): "[L'étendue] n'est qu'un Phénomène, une abstraction de plusieurs choses réelles." The former passage is an insertion of the 1742 edition.

[26] Du Châtelet (1740, 149): "Ainsi, si nous pouvions voir distinctement tout ce qui compose l'étendue, cette apparence d'étendue, qui tombe sous nos sens, disparoîtroit; & notre Ame n'appercevroit que les Etres simples éxistans les uns hors des autres, de même que si nous distinguons toutes les petites portions de matiére differemment mues, qui composent un portrait, ce portrait qui n'est qu'un Phenoméne disparoîtroit pour nous."

Indeed, this view is already present in an earlier part of the *Institutions*, namely in Chap. 5, where Du Châtelet makes her case for a relativist and plenist account of space. She begins by remarking that it "will not be useless to examine how we come to form the ideas of extension, space and the continuum."[27] She then argues that the idea of extension is an artefact of the way in which we represent objects to ourselves: When we imagine two distinct entities, it is natural for us to imagine them as spatially separated. In particular, we imagine anything that is distinct from ourselves as outside ourselves. As a consequence of this, we cannot imagine a union of two distinct things without this image of two things making one being accompanied by another notion, and this accompanying notion is what we call extension. The ideas of space and the continuum are then derived from that of extension.

Wolff's explanation of the phenomenality of extension is very similar, as Stan has remarked (2018, 483). In the *Cosmology*, Wolff writes that "extension and continuity are phenomena" and is keen to remark that his position is different from that of the idealists, who "call a phenomenon that which merely appears to exist, but which has no reality outside the mind."[28] Like Du Châtelet, he explains the phenomenality of extension with a visual example: Just as color is a confused perception of the way in which light is reflected by the surfaces of certain particles, so continuous extension is a confused perception of simple substances (Wolff, 1737, note to §229).

Though they hold that extended bodies have a foundation in reality, both Wolff and Du Châtelet are therefore idealists about continuous extension and about the properties examined by geometry generally. According to them, all that exists can be reduced to simple beings or atoms of nature and their aggregates, neither of which are infinitely divisible. The simple beings are unextended and therefore absolutely indivisible; the primary particles are naturally indivisible, but could still be divided into their constituent simple beings by God's power; the derivative particles, lastly, can become actually divided in the course of nature. Infinite divisibility is therefore a phenomenon in two senses: Firstly, it is natural to believe that extended bodies are infinitely divisible because we encounter no evidence to the contrary, but this is simply due to the crudeness of our perception and our tools. Secondly, there is indeed no limit to the possible divisions of the sort of extension that is treated in geometry, but that extension itself exists only as an abstraction from our confused perceptions of bodies. This is why it is possible to have at least two perspectives on the divisibility of a given body, as Du Châtelet explains:

> [F]or we have seen that it is proven by this principle [the PSR] that physical extension is ultimately composed of simple beings, and that consequently its divisions, even the possible ones, have real and positive limits. [...] So the infinite divisibility of extension is at the same time a geometrical truth and a physical error. [...] For any finite whole, for example a [body one] foot [long], is a composite of the finite and the infinite: this foot is finite, insofar as it contains only a certain number of simple beings, but I may suppose it to be divided into an

[27] Du Châtelet (1740, 96): "Il ne sera pas inutile d'éxaminer ici comment nous venons à nous former les idées de l'étenduë, de l'Espace, & du continu."

[28] Wolff (1737, §226): "Extensio & continuitas phenomena sunt. [...] Idealistae alio sensu phaenomenon appellant id, quod tantum existere apparet, nihil vero realitatis extra mentam habet."

> infinity [of parts], or rather into an unfinished quantity of parts, by considering that foot as an abstract extension.[29]

According to Euler, as we have already seen, the main mistake of the monadists is that they make an incoherent distinction between geometrical and physical extension. We are now in a position to see that his critique is misguided, even though the reconstruction of the monadist position that it is based on is fundamentally correct. His argument, which is found in letters VII and IX, goes as follows: They "allege that geometrical extension is an abstract being, from the properties of which it is impossible to draw any conclusions, with respect to real objects" (Euler, 1802, II, 26). This implies an absurd view of the relation between concrete objects and abstract notions, according to which a general notion can be truly predicated of an object that does not have all of the properties that are implied by the definition of that notion. Such a view, as Euler points out, would invalidate all scientific reasoning (Letter VII). Trying nevertheless to make their view plausible, the monadists according to Euler claim that "it is only an appearance of extension which is perceived in bodies, but that real extension by no means belongs to them." But this is merely a "wretched cavil," because it amounts to saying that there are two kinds of extension, one for bodies and one for geometry (Letter IX, 33–34).

It should be clear from the passages in Du Châtelet and in Wolff examined in this section, however, that neither of them would agree with this characterization: Euler seems to assume that if the kind of extension found in bodies is a phenomenon, there must still be a second, real extension as well. But this is not what Wolff or Du Châtelet mean: For them, the extension that is predicated of extended bodies is not radically different from the one that is the object of geometry. The only difference between the two notions is that in individual perceptions, extension is an epiphenomenon of a large number of simple beings, confusedly perceived, whereas the extension of geometry is an abstraction from many such impressions.[30] In other words, contrary to how Euler portrays their position, the monadists agree that physical bodies, insofar as they are extended, instantiate the general properties treated in geometry. What they disagree with is his assumption that extension is an irreducible primary quality and that it is therefore impossible to explain extended bodies in terms of something unextended. Indeed, some scholars of Wolff have explained his treatment of extension by a comparison with Locke: Extension for Wolff is a secondary quality, just as color and heat are secondary qualities for Locke (Hettche & Dyck, 2019, Sect. 5.1).

At the basis of Euler's disagreement with the monadists, therefore, are fundamental differences in their respective views on the ontology of extension. For Euler,

---

[29] Du Châtelet (1740, 183–85): "[...] car on a vû qu'il est prouvé par ce principle, que l'Êtendue Physique est à la fin composée d'Etres simples, & que par conséquent ses divisions, même possibles, ont des bornes positives & réelles. [...] Ainsi, la divisibilité de l'étendue à l'infini est en même tems une verité Géometrique, & une erreur physique [...] Car un tout fini quelconque, un pied, par example, est un composé de fini, & infini: ce pied est fini, en tant qu'il ne contient qu'un certain nombre d'Etres simples, mais je puis le supposer divisé en un infinité, ou plûtôt en une quantité non finie de parties, en considérant ce pied comme une étendue abstraite."

[30] For the question whether Du Châtelet is a realist or an idealist about extension, see also (Stan, 2018, 493).

the only adequate explanation of the fact that human experience of the external world includes extension is that extended things exist. Consequently, the very fact that the simple beings postulated by the monadists have no extension means that they are inadequate to explain extension. This is why Euler complains that "parts destitute of extension are incapable of producing a real extension; it can be, at most, an appearance, or a phantom which dazzles by a fallacious idea of extension" (1802, II, 58).

8.6 Two Proofs of Extension

Neither Du Châtelet nor Wolff would have agreed that their view makes extension into a "phantom" or that the perception of an assembly of simple beings produces merely a "fallacious idea" of extension. Their point, rather, is that extension *is* nothing but the idea which accompanies the confused perception of a multitude of simples. Whereas to Euler, the claim that extension arises from unextended simple beings is a simple category error, from the monadist point of view it is no more impossible than the production of color from properties that are not color.

However, explaining why a thesis is not immediately contradictory is not the same thing as showing that it is true. In this last section, I shall examine what positive reasons Wolff and Du Châtelet give for thinking that their view of extension is true or at least more plausible than the alternative (i.e., extension as a primitive property that cannot be further explained). It seems to me that there are two general argumentative strategies open to them. The first line of argument is this: Since it is certain, for independent metaphysical reasons, that simple substances are all that exists strictly speaking, they must be the cause of the phenomena of extension, even though it may be impossible for the human mind to understand how exactly that comes about. The second kind of argument would consist in giving an account of how the phenomena of extension can be explained in terms of conjunctions of simple beings. In this second case, the existence of the simple substances is a conclusion rather than a premise: Simple substances are inferred as the best explanation for extension, rather than the phenomenality of extension being inferred from the existence of the simple substances. This is the argumentative strategy that is usually employed when one wants to argue for the existence of insensible metaphysical entities, i.e. showing that the best available explanation for some uncontroversial fact is only possible if the entities in question exist. In order for this to work, however, one must present a cohesive story of how the features of the hypothetical metaphysical entities explain the observed features.

In the case of Du Châtelet, there are two passages in particular that fit well with this second strategy: Firstly, the explicit reason that she gives for introducing simple beings in the first place in §120 is that something without extension is needed to serve as the sufficient reason of extension. Secondly, in §133, she gives an argument that seems designed to show precisely what is needed for an inference to the best explanation, namely that simple beings can sufficiently explain continuous extension.

She begins the paragraph by noting that our perceptions of the world in general arise from confused perceptions of the many simple beings, each of which has its individuating properties but stands in relations to all others. She then continues:

> It is also through this dissimilarity [of the elements] that it is possible to understand how non-extended beings can form extended beings; and how the sufficient reason of the extension of matter is found in the simple beings, for the elements exist necessarily outside one another (since one can never be the other), and since they are all united with and linked to one another, as we have just seen, there results a compilation of multiple diverse beings that all exist outside one another and which are one whole in virtue of their connection; but I have shown that we cannot represent extension to ourselves in any other way than as a compilation of diverse coexisting things that exist outside one another; and so, the Leibnizians conclude that an aggregate of simple beings must be extended.[31]

There are three key features of the aggregated simple beings that Du Châtelet here singles out as relevant: Each simple being is different from every other in virtue of its specific internal properties, the simple beings are outside one another, and they form a whole. These three features together, so the argument goes, explain the phenomena of continuous extension. However, as Brading has previously remarked, there is a further premise in Du Châtelet's argument, namely that "we necessarily represent a multiplicity as spatially extended" (Brading, 2018, 158). I would argue that there are two ways to understand the necessity involved in this premise, or rather, how this necessity can be known. The first reading is suggested by the earlier passage that Du Châtelet refers to with the phrase "I have shown," which is found in §77. As already mentioned above in Sect. 8.5, she explains that the notion of continuous extension necessarily accompanies any mental representation of mutually external but linked entities. The phrasing in the summary in §133 is that extension can only be represented as a compilation of diverse coexisting things, but strictly speaking, the argument of §77 is that we cannot represent diverse coexisting things except as extended. The thesis therefore seems to be that none of these two types of representation can occur without the other, that they are necessarily correlated.

The most natural way of reading §77, in my view, is that one can find out by introspection that the two notions must always occur together, i.e., that their necessary connection is a purely psychological fact: Du Châtelet writes that "we feel that, when we consider two things as different and distinguish one from the other, we place one outside the other in our mind."[32] But if this is the claim, the justification

[31] Du Châtelet (1740, 147–48): "C'est encore par cette dissemblance que l'on peut comprendre comment les Etres non étendus peuvent former des Etres étendus; & que l'on peut voir comment la raison suffisante de l'étendue de la matière se trouve dans les êtres simples, car les Eléments éxistent tous nécessairement les uns hors des autres, (puisque l'un ne peut jamais être l'autre), & étant tous, comme on vient de le voir, unis & liés ensemble, il en résulte un assemblage de plusieurs Etres divers, que éxistent tous les uns hors des autres, & qui par leur liaison sont un tout; mais j'ai fait voir que nous ne pouvons nous représenter l'étendue que comme l'assemblage de plusieurs choses diverses coéxistentes, & qui éxistent les unes hors des autres (§77): donc, concluent les Leibnitiens, un aggrégat d'Etres simples doit être étendu."

[32] Du Châtelet (1740, 96–97): "Nous sentons que, lorsque nous considérons deux choses comme différentes, & que nous les distinguerons l'une de l'autre, nous les plaçons dans notre esprit l'une hors de l'autre." The paragraph is based on (Wolff, 1736, §544).

that Du Châtelet offers for it is hardly enough to convince someone with different intuitions. The justification consists in a series of examples, all of which are taken from the realm of extended bodies: In the first case, one imagines a house that one has never seen before, in the other, two duplicates of the same man. It is true that one tends to imagine these things as outside one another, but that is because, being extended bodies, they must exist according to the rules of three-dimensional space. That is arguably not the case with abstract examples like purely geometrical figures, numbers, or pure concepts, e.g. "to travel." This does not immediately disprove the claim, since it is about "things" in a strong sense, i.e., things that have their own essence, which such abstractions do not. But the key class of examples is precisely the one type of substance that cannot be assumed to play by the rules of space, namely, simple beings. Insofar as it is possible at all to imagine two simple, unextended, mutually connected substances with distinct essences (as opposed to merely holding that concept in the mind) it seems very well possible to also imagine that their connection is of an entirely different kind than that of two bodies in space. Hence, while it is *possible* that in the hypothetical world where simple beings form the basis of reality, confused perceptions of their aggregates would bring about the real phenomena called extended bodies, they could just as well not cause that kind of phenomenon. Unless there is some other reason to believe that simple beings must be perceived as mutually external, therefore, the power of the simple beings to explain extension is questionable. But since simple beings are nothing but a hypothesis introduced in order to explain extension, the entire argument for their existence is put into question as well.

Might it be that we know of the necessary correlation of the notions of extension and of mutually connected beings through conceptual analysis and not through introspection? This is how the parallel argument in §§215–222 of Wolff's *Cosmology* has often been read. Wolff argues, like Du Châtelet, that the simple beings (or simple substances, in his terminology) are distinct. Also like her, he argues that since they are mutually distinct, they are outside one another (§219) and united in virtue of their internal properties, since they "contain the reason why they coexist in this way rather than another" (§220).[33] He then draws the conclusion that "since that in which diverse [things], that is, [things] existing outside one another, are united must be extended, the aggregates of the elements are extended" (§221).[34] In §222, he adds that the simple elements must not only be outside one another, but that they must even form a continuum. He argues that since no individual element can be substituted for another without changing the identity of the whole aggregate, there is an "order of coexistence" of any given element with all others, i.e., it is impossible for a new element to be inserted between two existing ones, which is the definition of a continuous arrangement. As is often the case with Wolff, the writing is dry and logical,

[33] Wolff (1737, §220): "In ipsis enim elementis continetur ratio, cur hoc modo potius coëxistant, quam alio."

[34] Wolff (1737, §221): "Aggregata elementorum extensa sunt. Dum enim elementa aggregantur, extra se invicem existunt (§219) ac inter se uniuntur (§220). [...] Quare cum extensum esse debeat, in quo multa diversa, seu extra se invicem existentia uniuntur; aggregate elementorum extensa sunt."

making it appear as if the conclusion follows with necessity from the definitions of the notions involved. One needs to follow up the references to the *Ontology* to find out that he, too, defines extension in psychological terms: "If we represent multiple things that are distinct and exist outside one another as if they were one, the notion of extension arises" (§548).[35]

As multiple commentators have pointed out, if we take Wolff at his word and read his argument as being that the notion of continuous extension necessarily follows from the notion of distinct but linked entities, it seems to have not one but two holes: Firstly, it is not clear why distinctness would imply being spatially outside one another, and secondly, it is also not clear why a collection of entities, even if they are outside one another, would necessarily constitute a *continuous* extension. Arthur gives numbers and successive perceptions as counter-examples for both steps. Concerning the latter claim, he sees the central problem in the fact that for Wolff's demonstration to work, the specific order of the elements in the aggregate need to be derived from the properties of the simple substances (2018, 30–31). More generally, De Risi's diagnosis (2007, 305–307) is that while Leibniz was successful in proving that all the properties of geometrical space can be derived from the relational concept of *situs*, the general assertion (made by Wolff and others defending physical monads) that extension can be reduced to relations between simple beings simplifies away too much of Leibniz' mathematical analysis to still be convincing. To Rutherford, too, Wolff's argument is "a simple conceptual error: the conflation of the idea of non-identity with that of externality" (2004, 323).

One (drastic) way of dealing with these difficulties is to read the main argument in the opposite way. In §133, as we have seen, Du Châtelet refers back to §77, where she had first introduced extension as the notion which arises when we imagine multiple distinct entities that are united in some way. This might be read to mean that in her view, it is not possible for us humans to know how exactly the notion of extension is produced. This in turn would imply that she is following not the second, but the first of the two types of general strategies outlined above. The goal of the argument would then not be to prove the existence of simple beings, but rather the phenomenality of (continuous) extension.

Though it evades the objections leveled against the traditional reading of Wolff, however, this interpretation faces its own problems, both interpretative and argumentative. The interpretative troubles are the following: Firstly, if the role of the simple beings in constituting the continuum can only be known after their existence is already secured, a separate argument for their existence is needed. One would therefore expect Du Châtelet to give such an argument, for example a version of Leibniz's claim that simple substances are needed to explain substantial unity. As we have seen however, Du Châtelet does not proceed in this way, and rather introduces simple beings as the sufficient reason for the existence of divisible bodies in §121. Secondly, the idea that (continuous) extension cannot be completely understood by the human mind according to Du Châtelet is difficult to reconcile with the passages

[35] Wolff (1736, §548): "Si plura diversa, adeoque extra se invicem existentia, tanquam in uno nobis repraesentamus; notio extensionis oritur."

in which she appears to explain how precisely that is possible. The most obvious such case is §133, where the stated aim is to show "how the sufficient reason of the extension of matter is found in the simple beings" (Du Châtelet, 1740, 147). In the earlier chapter on the nature of space, she also gives an explanation why space and the bodies in it must necessarily appear to us not only as extended, but as continuous (1740, 102–103).

Apart from these textual issues, however, the problem with the alternative reading is that it fails to fill the phenomenalism of the monadists with meaning. If it is admitted that there can be no explanation of how the impressions made by a multitude of unextended simples create the phenomena of continuous extension, the result would be "not phenomenalism, but just a philosophical defeat" (De Risi, 2007, 307). If the alternative is this kind of empty reduction, it would indeed be preferable to regard extension as an irreducible attribute of body.

8.7 Conclusion

In the Sect. 8.1 of this paper, I briefly discussed the Leibnizian origins of the labyrinth of the continuum and showed that the expression occurs in both Wolff and Du Châtelet. I also introduced Euler as a prominent critic of the "monadist" position on the divisibility of matter. The rest of the paper was structured along two main parts of Euler's critique. His first accusation was that the indivisibilist position arises from a confusion between two sorts of divisibility, namely divisibility by natural means, and divisibility in principle. In Sect. 8.3, I showed that contrary to Euler's suggestion, this sort of distinction had been central in the debates on matter theory at least since Descartes. In Sect. 8.4, I showed that the two sorts of divisibility both have their place in Du Châtelet's system, namely in the fundamental and the derivative particles (which have their direct origin in Wolff). As I noted, Du Châtelet made extensive changes to the chapter in which she discusses the question of primary and secondary particles, though the revised version of the chapter contains substantially the same position as the original one, presented more clearly.

Section 8.5 of the paper turned to the main point of Euler's critique, i.e., that in order to accommodate the infinite divisibility of geometrical objects, the monadists make an incoherent distinction between geometrical and physical extension. This turned out to be based on a misrepresentation of the monadist position, since the phenomenalism of Du Châtelet and Wolff does not mean that there are two sorts of extension. They hold, rather, that bodies, insofar as they are extended, are merely phenomena, so that the realm of geometry as such is merely phenomenal as well. I have argued that this is a perfectly adequate response to Euler's arguments, and that the reason why the monadist position seems so absurd to Euler is that they do not share his assumption that extension is a real and irreducible property of bodies. Wolff and Du Châtelet therefore present a coherent solution to the perceived problems of the continuum. As discussed in Sect. 8.6, however, that is not the same as giving good arguments for why one should prefer the monadist explanation of bodily phenomena to one that simply posits extension as an irreducible property. The goal is to explain how continuous bodies, as they are part of our phenomenal experience, are produced by the simple beings. The fundamental problem, however, is that we simply know too little about them for such a derivation. This is why both Du Châtelet and Wolff ultimately cannot fulfill the promise implied by their claim that the simple beings are the sufficient reason for extension. Euler, in his polemic attempt to paint the very idea of unextended simples as incoherent, misrepresents the relationship of simples and composite bodies in such a way that he fails to notice this more subtle but ultimately more serious weakness of physical monadism.

On most of the topics treated in this paper, Du Châtelet's stance and arguments have turned out to be quite close to those of Wolff. Their most significant difference lies in how they present the central premise that we necessarily represent a multitude of connected things as extended: Whereas Wolff's demonstration seems designed to prove a conceptual necessity, Du Châtelet's language is psychological. Her premise seems to be about the way in which our mental representation works, namely that it is impossible to imagine two things that are distinct in essence but not outside one another. This fits in with her general approach in chapter five of the *Institutions* (§§77–93), which is framed as a description of how we come to have the concept of space and related notions. Wolff might well have agreed with this description of the human imagination, just as Du Châtelet might have believed that the necessary correlation of the notions of extension and of distinct but connected beings is founded on a metaphysical necessity. But the derivation of extension from simple beings that Wolff puts forward in the *Cosmology* makes no direct mention of the fact that it is grounded in assumptions about the working of the faculty of representation—compare (Wolff, 1737, §§219–223) with (Wolff, 1736, §§544 and 548). Du Châtelet, in contrast, signals from the beginning that her starting point are the notions of extension, space and body as they are in the mind, which makes her version of the argument clearer and arguably stronger in comparison.

References

Arthur, R. T. W. (1998) Infinite aggregates and phenomenal wholes: Leibniz's theory of substance as a solution to the continuum problem. *Leibniz Society Review 8,* 25–45.

Arthur, R. T. W. (2018). *Monads, composition, and force: Ariadnean threads through Leibniz's Labyrinth* (1st ed.). Oxford University Press.

Brading, K. (2018). Émilie Du Châtelet and the problem of bodies. In E. Thomas (Ed.), *Early modern women on metaphysics* (pp. 150–168). Cambridge University Press.

Boyle, R. (1772). The works of the Honourable Robert Boyle (T. Birch, Ed.; 1–6). Rivington.

Calinger, R. (1976). Euler's 'letters to a princess of Germany' as an expression of his mature scientific outlook. *Archive for History of Exact Sciences, 15*(3), 211–233.

De Risi, V. (2007). Geometry and monadology: Leibniz's analysis situs and philosophy of space. In *Science networks. Historical studies* (Vol. 33). Birkhäuser.

Descartes, R. (1897). In P. Tannery, & C. Adam (Eds.) *Oeuvres*. Léopold Cerf.

Du Châtelet, E. (1740). *Institutions de physique*. Prault.

Du Châtelet, E. (1742). *Institutions de physique* (2nd ed.). Aux depens de la Compagnie.

Euler, L. (1768). *Lettres à une princesse d'Allemagne sur divers sujets de physique & de philosophie*. Académie Impériale des Sciences.

Euler, L. (1802). *Letters of Euler to a German princess, on different subjects in physics and philosophy* (H. Hunter, Trans.) (2nd ed.).

Euler, L. (1963). *ПИСЬМА к УЧЕНЫМ (letters to scholars)*. Acadamy of Sciences of the USSR.

Garber, D. (1992). Science and its conceptual foundations. In *Descartes' metaphysical physics*. University of Chicago Press.

Garber, D. (2009). *Leibniz: Body, substance, Monad*. Oxford University Press.

Hettche, M., & Dyck, C. (2019). Christian Wolff. In E. N. Zalta (Ed.) *The stanford encyclopedia of philosophy*. Metaphysics Research Lab, Stanford University.

Hutton, S. (2012). Between Newton and Leibniz: Emilie du Châtelet and Samuel Clarke. In R. Hagengruber, (Eds.), *Emilie du Châtelet between Leibniz and Newton* (pp. 77–95). Springer Netherlands, Dordrecht.

Janiak, A. (2018). Émilie Du Châtelet: Physics, Metaphysics and the case of gravity. In E. Thomas (Ed.), *Early modern women on metaphysics* (pp. 49–71). Cambridge University Press. https://doi.org/10.1017/9781316827192.004

Leibniz, G. W. (1720). *Essais de théodicée sur la bonté de Dieu, la liberté de l'homme, et l'origine du mal*. David Mortier.

Leibniz, G. W. (1875). In C. I. Gerhardt (Ed.) *Die Philosophischen Schriften von Gottfried Wilhelm Leibniz*. Weidmann.

Leibniz, G. W. (1970). In L.-F. Münster (Ed.) *Sämtliche Schriften und Briefe. Reihe 6: Philosophische Schriften* (Vol. 6, pp. 1672–1676). Akademie Verlag.

Leibniz, G. W. (1976). In L. E. Loemker (Ed.) *Philosophical papers and letters: A selection* (L. E. Loemker, Trans.). Riedel.

Levey, S. (1998). Leibniz on mathematics and the actually infinite division of matter. *The Philosophical Review, 107*(1), 49. https://doi.org/10.2307/2998315

Phemister, P. (2005). The New Synthese Historical Library. In *Leibniz and the natural world: Activity, passivity, and corporeal substances in Leibniz's philosophy* (Vol. 58). Springer.

Rohault, J. (1723). *Rohault's system of natural philosophy illustrated with Dr. Samuel Clarke's notes taken mostly out of Sir Isaac Newton's philosophy. ... Done into English by John Clarke*. Knapton.

Rutherford, D. (2004). Idealism declined: Leibniz and Christian Wolff. In P. Lodge (Ed.), *Leibniz and his correspondents* (pp. 214–237). Cambridge University Press.

Stan, M. (2018). Emilie Du Châtelet's metaphysics of substance. *Journal of the History of Philosophy, 56*(3), 477–496. https://doi.org/10.1353/hph.2018.0048

Suisky, D. (2012). Leonhard Euler and Emilie du Châtelet. On the Post-Newtonian development of mechanics. In R. Hagengruber, (Eds.), *Emilie du Châtelet between Leibniz and Newton* (pp. 113–155). Springer Netherlands, Dordrecht.

Wilson, C. (1989). Studies in Intellectual History and the History of Philosophy. In *Leibniz's metaphysics: A historical and comparative study*. Manchester university press.

Wolff, C. (1736). *Philosophia prima sive ontologia: Methodo scientifica pertractata, Qua Omnis cognitionis humanae principia continentur*. Editio nova, Priori emendatior. Renger.

Wolff, C. (1737). *Cosmologia generalis, methodo scientifica pertracta, qua ad solidam, inprimis dei atque naturae cognitionem via sternitur*. Editio nova priori emendatior. Renger.

Gabriel Müller is an employee at the University Library Basel. His research focuses primarily on Anti-Aristotelianism in the seventeenth century.

Chapter 9
Émilie Du Châtelet on Freedom: The Wolffian Context

Jan J. Forsman and Jani I. Hakkarainen

9.1 Introduction

Émilie Du Châtelet's (1706–1749) *On Freedom* (c. 1737; henceforth OF) has in recent years received growing interest from scholars (e.g., Jorati, 2019; Lascano, 2021, forthcoming; Wells, 2022). Much attention has been paid to whether her account of freedom is an instance of compatibilism or libertarianism, particularly in relation to its influence from Leibnizian and Clarkean accounts.

We take a different approach to the text and contextualize Du Châtelet's account of freedom with Christian Wolff's (1679–1754) theory of free will and action, especially in his work the *German Metaphysics* (1719; henceforth GEM).[1] According to Wolff, all actions must have a ground (what the intellect considers good) and for those actions to be free, they must have their ground internal to the soul (instead of due to an external ground); if a ground is present and is sufficient to move the will, an action follows with moral necessity.

Wolff's account resulted in a conflict with the German Pietists who argued that emphasizing an internal ground for freedom would reduce human action to the mechanistic workings of a clock. This led to a debate on whether internal spontaneity is central for free actions on its own, or whether there needs to be some causal contribution from the will, independent of reason, to the action for that action to be free.

[1] The original title is *Vernünftige Gedanken von Gott, der Welt und der Seele des Menschen, auch allen Dingen überhaupt, den Liebhabern der Wahrheit mitgetheilet.*

J. J. Forsman (✉)
University of Iowa, Iowa City, IA, USA
e-mail: jan-forsman@uiowa.edu

J. I. Hakkarainen
Tampere University, Tampere, Finland
e-mail: jani.hakkarainen@tuni.fi

C. Carus and J. K. McDonough (eds.), *Émilie Du Châtelet in Relation to Leibniz and Wolff*, Women in the History of Philosophy and Sciences 23,
https://doi.org/10.1007/978-3-031-80541-7_9

Contextualizing Du Châtelet's account with Wolff's *German Metaphysics* and the following Pietist debate helps in clarifying certain puzzling parts of *On Freedom*, particularly ones that have to do with the connection of the will to physical action and the issue of whether Du Châtelet supports a more compatibilist or libertarian view of freedom. Regarding the latter, we propose a conciliatory reading due to textual evidence for both compatibilist and libertarian accounts. We argue that Du Châtelet does not commit to a clear compatibilist or libertarian position in *On Freedom* and merely responds to what she calls the "illustrious opponents" of freedom in the text (OF 485).

The structure of the paper is as follows: In the second section, we discuss *On Freedom* and problematize the dispute over whether it supports compatibilist or libertarian understanding of freedom. The third section is devoted to a brief exposition of Wolff's account of free will and action. We focus on his account as it is presented in the *German Metaphysics* since we present evidence that this was the work by Wolff that Du Châtelet could have been reading while composing *On Freedom.* As the main topic of this paper is Du Châtelet, we do not discuss the background of Wolff, in Leibniz for instance, although it is obvious that regarding free will Wolff is heavily indebted to Leibniz (even though there are essential differences). In the fourth section, we introduce the controversy between Wolff and the Pietists concerning freedom in 1720's and 30's. This leads us to argue in the fifth section that Du Châtelet was likely acquainted with Wolff's account of free will, as well as his subsequent debate with the Pietists, when she wrote *On Freedom.* In the sixth section, we argue that this contextulization helps us to understand Du Châtelet's own thinking about freedom.

9.2 *On Freedom*

On Freedom is an unpublished manuscript that Voltaire (1684–1778) sent to Crown Prince Frederick (1712–1786), who was to later become Frederick the Great, in late-October 1737.[2] While Voltaire sent the manuscript under his own name, later research has indicated a sizable, perhaps a sole role from Du Châtelet in its composition. L. G. Janik has made the case that the manuscript was intended to be included in

[2] We rely on both a published version of the original text in *Oeuvres complètes de Voltaire*, vol. 14, 484–502 (Barber 1989), and a scribal copy with additions and corrections preserved in the National Library of Russia in St. Petersburg, with an Electronic edition provided by the Center for the History of Women Philosophers and Scientists, University of Paderborn, in cooperation with the Centre international d'étude du XVIIIe siècle, Ferney-Voltaire (referred to as St. P. ms.). For the translation, we use Jorati's (2021).

The Foundations of Physics (1740) as Chap. 5 of the work (Janik, 1982, 88).[3] The manuscript, however, was never published by either Voltaire or Du Châtelet.[4]

In recent years, the manuscript has garnered new interest, especially relating to scholarly debate on what kind of freedom Du Châtelet supports in the text. Jorati (2019) takes *On Freedom* to agree with Leibnizian *compatibilism*. Wells (2022) supports a reading according to which *On Freedom* follows more closely Clarkean *libertarianism*. And Marcy Lascano (forthcoming) has recently argued that Du Châtelet's understanding of freedom in the text is Lockean.

On Freedom is indeed a challenging text, and it includes several perplexing passages. One of the puzzling parts of the text is that it appears to include support for both compatibilist and libertarian readings about the relationship between the intellect and the will. A supporter of the compatibilist reading can find support in the following passages:

> Thus, due to his intelligence, man is under the necessity of willing that which his judgment presents to him as best. (OF 496)
>
> To will that which does not please is a genuine contradiction; and being free means doing what one judges to be best, what brings pleasure. (OF 496)

In passages like these, Du Châtelet seems to insinuate that freedom is simply the case of following our judgment and doing what it tells us to be the best course of action.

However, a supporter of libertarian reading can find support in her reading in Du Châtelet's discussion of a case where the soul chooses between two perfectly similar things, which we call th*e Buridan's ass passage:*

> [W]hat is quite certain is that the self-moving power, the sole and true source of freedom, cannot be destroyed by the indiscernibility of two objects; for insofar man has this self-moving power, man is free. (OF 493)

Here, Du Châtelet seems to conclude that freedom involves having the physical power to act one way or another in any given situation.

Another passage, that has perplexed scholars, is Du Châtelet's discussion of the connection between the will and physical action, where she mentions that "the will is never the cause of our actions, even though it is their *occasion*; for an abstract notion cannot have any physical influence over the physical self-moving power" (OF 493; emphasis added). Some accounts attempt to explain the passage by referring to Samuel Clarke's (1675–1729) usage of *occasional causation* (e.g., Jorati, 2019), which Du Châtelet is known to have been aware of around the time from the Leibniz-Clarke correspondence (Letter to Prault, July 16, 1738: Du Châtelet, 2018, I:367).

---

[3] The St. P. ms. has the heading Chapitre 5 and includes a note on the last page "sur la liberté par m^{e} du chastelet". The additions and corrections are furthermore written only in Du Châtelet's handwriting. Note, however, that parts of the text follow closely Chap. 7 of Voltaire's *Traité de metaphysique* from three years earlier (Barber 1989, 460–467; see also Pattersson 1938). This leads us to consider that the work, as well as the previous *Traité*, might have been a joint effort by the two. Nevertheless, we follow the convention of attributing the manuscript solely to Du Châtelet.

[4] *Traité de metapysique* was never published either.

We believe that contextualizing *On Freedom* with Wolff, instead of Leibniz, Clarke or Locke, can clarify certain problematic parts of the text. Particularly, Wolff's account of freedom can illuminate Du Châtelet's discussion of the will's connection to physical action, as well as the debate over whether she should be read as supporting more strongly compatibilism or libertarianism in the text.[5]

Before we get further in analyzing Du Châtelet's text, however, we consider it important to offer a short summary of Wolff's account of free will in the *German Metaphysics*, and of the philosophical and political scandal that followed from this view, known as the Pietist debate.

9.3 Wolff on Free Will in the *German Metaphysics*

Imagine being in a bookshop and considering whether to buy *The Foundations of Physics* by Du Châtelet or *Monadology* by Leibniz. Let us suppose that by your understanding (*Verstand*) you represent distinctly the *Foundations* as good and better than *Monadology* to yourself. The distinctness of this representation means that you see in what the superiority of the *Foundations* to *Monadology* consists. Then, according to Wolff, you *must* want and choose the *Foundations*. Your understanding representing it distinctly as good and better is a *motive* and *ground* necessitating your will in the situation in Wolff's view. Within the constraints of this distinct representation (*Vorstellung*), you do not have an option to will and choose otherwise (see below for a qualification). "[W]e will nothing except what we take to be good" (§506).[6] Yet Wolff thinks that your choice and willing are free; determination of the will by the understanding does not undermine freedom of the will. Actually, the freedom of the will consists partially in being determined by intelligible grounds that are motives for the will. Wolff is a compatibilist regarding the relation between the will and the understanding.

Original German terms might help understanding his position. "Ground" is of course "Grund" and "motive" is "Bewegungs-grund", which means literally ground for movement; it is Wolff's rendering of the Latin "motivus". The representation of good is ground for the will to move to the resolution, which is then grounded. Ground in general is that by which we understand (*Verstehen*) why the grounded is (§29).[7] By the representation of good, we understand why the will chooses what it does (§27). If I represent the *Foundations* as good, my wanting it is grounded in the representation. The soul, in turn, is the cause (*Ursache*) of the resolution by the will since the ground of the resolution is in the soul (§§29 and 518).

---

[5] We do not mean to say that Leibniz, Clarke, and Locke were not influential for her account, of course, just to extend the contextualization to include Wolff and to shed light on OF by this further contextualization.

[6] The references are to GEM [1719], translated by Dyck (2019), unless otherwise noted.

[7] "Grund" is Wolff's translation of the Latin "ratio" (GEM [1736], the 1st Register). Therefore, "Grund" may also be translated into English by "reason".

Wolff emphasizes that the understanding necessitates the will in terms of *moral necessity* only (*die Nothwendigkeit der Sitten*) (GEM [1736], §521).[8] It is still *possible in itself* (*an sich selbst*), that is, not absolutely necessary (*schlechterdinges nothwendig*)[9] that the will chooses otherwise than the understanding dictates (§§515–7 and 521). In the space of possibility in itself, I have freedom to will and choose otherwise independently from motives. For example, it is possible in itself that I choose *Monadology* even if my understanding morally necessitates my will to want the *Foundations*. Choosing *Monadology* does not contain anything contradictory (affirming and denying the same at the same time) within it, that is, in itself (§§11–2). The same holds not only for acts of the will but also for actions (§§515–6).

So far, we have discussed two of Wolff's *definiens* of free actions: distinct understanding and the denial of absolute necessity (§§514–5). A representation by the understanding is the ground and motive making it morally rather than absolutely necessary that the will chooses according to the motive. The will is free to choose otherwise than the motive determines in terms of the possibility in itself. Wolff's third and perhaps the most significant definiens is *spontaneity*. His German term here, "Willkühr" is his rendering of the Latin "spontaneitas" (GEM [1736], 1st Register) The soul has spontaneity insofar as it has the ground of its actions in itself (§518). Spontaneity means that the actions of the soul are intrinsically rather than extrinsically grounded, in a word self-grounded or self-determined.[10]

Accordingly, Wolff defines *freedom* as follows:

> [*F*]*reedom* is nothing other than the capacity of the soul, through [*durch*] its own spontaneity, for choosing [*Wehlen*] that one among two equally possible things that pleases it most. (§519; translation altered).

When we connect this definition to that of capacity (*Vermögen*, §117), we can say that freedom is the *capacity of choosing spontaneously*: the possibility of bringing about the resolution as a change grounded intrinsically in the soul. Generally, the source (*Quelle*) of change is power (*Krafft*, §115).[11] Therefore, the soul has the power to choose spontaneously. Wolff says: "For the representations that are needed for motives (§496) are in it and come from it, and it is inclined through its own power towards whatever pleases it" (§518). He even claims: "*Everything in the soul comes from it.*" (§754) Indeed, it is of utmost importance for Wolff later in Chap. 5 that the soul has a single power only (§745) since it is a simple thing (*Ding*) indivisible into parts of whatever kind (§742). If the soul had several powers, it would compromise the simplicity of the soul. The understanding, reason (*Vernunft*), the will and sensual desire (*die sinnliche Begierde*) are only different names for the only power of the soul (§747). They are not numerically distinct powers. It is the one power of the soul that

[8] To be precise, Wolff adopts the German term for "moral necessity" in the third edition of GEM in 1725.

[9] In his glossary, Wolff renders the Latin "Absolute necessarium, geometrice necessarium, metaphysice necessarium" by "Schlechterdinges nothwendig" (GEM [1736], the 1st Register).

[10] Later Wolff defines spontaneity as an intrinsic principle determining acting (Wolff 1732, §933).

[11] Wolff's spelling is indeed "Krafft". He distinguishes it from "Vermögen", capacity (§117).

produces diverse effects like understandings, reasonings, acts of the will and sensual desires (ibid.).

Among those diverse powers understanding is having a distinct thought or concept of a thing, that is, being able to tell what the difference is between that thing and other thought things (§§276 and 206). If I understand what the triangle is, I can describe the constitution of its difference from the square (§206). Reasonings, in turn, are inferences between propositions and reason is accordingly the "insight [*Einsicht*] which we have into the connection of truths", or the capacity of that insight (§368).[12]

The will is the inclination (*Neigung*) of the mind towards a thing (*Sache*) because of taking the thing as good (§492). When this inclination involves an *indistinct* representation of the goodness of the thing, we have a sensual desire (§434). By an indistinct representation, we cannot tell what constitutes its difference to other thought things (§214). According to Wolff, this happens, for instance, when we have representations of different colors. Clearly, they are different, but what is the constitution of the difference? (ibid.) When a sensual desire (or aversion) has "a notable degree", Wolff calls it a "sensual affect" (§439).

Wolff thus has three conditions for a free act of the will. (1) There is a ground for the act in a distinct representation of the understanding. Wolff specifies that the capacity of reason is the ground of freedom (§520). Reason establishes what is good since it provides insight into the connections of truths about what is good (ibid.). (2) The act is morally rather than absolutely necessary. (3) The act of the will is intrinsically grounded in the soul, that is, self-grounded (spontaneity). We may therefore label Wolff's account *intellectualist compatibilism* since free will is compatible with and consists partly in the rational determination by the understanding. Indeed, there has to be a sufficient ground for each act of the will in Wolff's view since he holds an unrestricted principle of sufficient reason (PSR, "*Satz des zureichenden Grundes*"): "*everything that is must have a sufficient reason why it is.*"[13] (§30) Due to the PSR, Wolff rejects *liberty of indifference*, that is, choosing without motives that would make one of the alternative options more preferable than the others (§511). He is committed to denying that Buridan's ass could choose one haystack over the other.

It needs to be noted though that this does not make Wolff overtly rationalist in his psychology. His account of *what actually motivates us* involves also sensual affects and hence sensual desires and aversions are motives for the will (§502). In choosing between ice cream flavors, for example, that chocolate tastes more pleasant than vanilla to me would be a motive for choosing chocolate over vanilla. It is only that in the free acts of the will, the motive comes from reason. Insofar as we are rational "masters" of our will rather than the will being a "slave" of sensual affects and desires, our will is free. Besides, Wolff discusses the case where we first prefer better but then something intervenes, and we end up willing and doing worse (§§517 and 521). It might happen, for instance, that I end up choosing and buying the *Monadology* instead of the *Foundations* even though I originally consider the latter better. The point is that *if* a motive is present, then it *morally* necessitates the will.

[12] Wolff adds the qualification about the capacity in the third edition (1725).

[13] At the least, the PSR is not restricted among actual beings.

Nevertheless, Wolff thinks that we should want the *true good* instead of the apparent (§424). Good is whatever "makes us and our condition more perfect" (§422) or ordered in a complete (*lauter*) manner (§156), which produces pleasure in us (§423). For example, health is good because it makes one's body (*Leib*) "more perfect": more ordered in a complete manner (§422). Being healthy is then pleasurable. Accordingly, "a *true good* is what yields a constant pleasure, or at least never causes displeasure" (§424). By reason, we have prospects to understand what is truly good.

As an upholder of intellectualist compatibilism, Wolff emphasis is on the freedom of the will. But he indicates that the freedom of *action* more or less follows from the freedom of the will since he argues for his definition of freedom by discussing free physical actions (§§513–8). Such free actions have a ground in reason and hence in the soul (§§518–20). This raises the question, what is Wolff's account of the relation between souls and bodies? Can there be efficient causation between them? By bodies (*Körper*) or corporeal things (*körperliche Dinge*), Wolff means composite (i.e., divisible) extended things that fill a space necessarily (§52).[14] Therefore, they differ essentially from souls, which are simple non-spatial things, as was seen above. In contrast to souls, bodies cannot think, nor it is possible that God could communicate thinking to them (§§738 and 741).

Could bodies nonetheless influence the thinking of the soul or could the will affect bodies in Wolff's view? His response is strongly negative. According to Wolff, bodies have the motor force (*vis motrix*) to communicate motion to other bodies and the force to resist motion (*vis inertiae*) (§607; Hettche & Dyck, 2019, Sec. 5.2). From these, he (§608, transl. in Watkins, 2009) concludes that "no body can move itself, and accordingly it must have an external cause, if it is to be moved (§30, 29)." But that external cause cannot be the soul. Wolff considers only one option here: natural influx (*Einfluss*) that would roughly be the transfer of attributes between souls and bodies. He rejects it for two main reasons. First, we do not have experience (*Erfahrung*) of the natural influx between souls and bodies (§§529 and 536). Second this influx would violate the conservation of motor force in the world, that is, in the composite corporeal whole (§§762 and 550–1). If the movement of bodies transferred to souls, then motion and hence motor force would be lost in the world. In the case that the soul produced motion in bodies by merely willing it, motor force in the world would be increased (§§550–1).

Wolff equally discards *occasionalism*, which he attributes to Descartes, on the grounds that occasionalism is committed to perpetual miracles (§§763–4; for the details, cf. below). Wolff thinks that the only option left is the pre-established harmony between souls and bodies (*die vorherbestimmte Harmonie*) (§765). Wolff's argument here is eliminative, but we should not assume that it is demonstrative. He seems to consider the pre-established harmony only as the most likely hypothesis among the three options (cf. §527).

[14] Here we do not need to take sides on the interpretative issue whether Wolff holds an idealist or a realist view of bodies in GEM. Under one reading, he does hold an idealist view in his later Latin works (Hettche & Dyck, 2019, Sec. 5.2).

We do not need to go into a detailed discussion of Wolff's hypothesis of the pre-established harmony in this occasion. It suffices to note that what happens in the soul agrees with what happens in bodies and that the PSR determines changes in both (§767). Therefore, the changes in both succeed each other "in a fixed order" (ibid.). Souls, on the one hand, and bodies, on the other, constitute determined orders. Wolff's hypothesis is determinist regarding both the soul and the body. God has instituted the agreement in the beginning, and everything follows from that by the PSR without His intervention (§§766–7). Wolff thinks that he, unlike occasionalists, is not committed to perpetual miracles. A further point to appreciate is that each soul is only internally determined after its first state instituted by God (§754). Wolff is not, however, a necessitarian since he thinks that there are counterfactual absolute possibilities.

As was seen above, the determinism of the pre-established harmony does not threaten human free will in Wolff's view. Let us suppose that my willing to raise my hand is grounded in my distinct understanding that it is good to ask a question. It is not absolutely necessary for me to want to raise my hand nor to raise it. Wanting to keep it down and not raising it are equally possible in themselves. In virtue of my distinct understanding, I choose to raise my hand. Then my hand arises without my willing to raise it naturally influencing its movement. My willing and the movement of my hand are just in harmony and there are sufficient grounds for both. I can also represent the movement of my hand by my soul since the soul has "a power of representing the world according to the position of its body [*Körper*] in the world." (§753).

In connection to Du Châtelet's account of freedom, it is noticeable that in spite of his rejection of what he considers *Cartesian* occasionalism Wolff says a few times that changes in the bodily organs of sense occasion (*veranlassen*) sensations of the soul (e.g., §§220 and 749). This shows that even a proponent of the pre-established harmony may say that changes in the bodies are occasions (*Veranlassung*) of changes in the soul without being committed to any kind of occasionalism.[15] Most likely, this just means something like that changes in bodies are harmonious happenings for changes in the soul.

9.4 The Boy Who Cried Wolff: Pietist Critique of Wolff on Free Will

In connection with free will and the pre-established harmony, Wolff was subject to a fierce attack by the Pietists, who were spiritualist reformers of German Protestantism in early modern times. It all started after Wolff had given his vice-rector farewell speech to an audience of more than 1000 people at the University of Halle in 1721. At Halle, the Pietist theologians August Hermann Francke (1670–1733) and especially Joachim Lange (1670–1744) were infuriated by Wolff's suggestion

[15] At the very least, this is how Dyck (2019) translates "veranlassen" and "Veranlassung" in this context.

that divine (Christian) revelation is not necessary for finding moral truths. In this context, Wolff spoke about Confucianism in positive terms. The Pietists were able to convince the King in Prussia, Frederick William I (1688–1740), who had the reputation of being a soldier king, of the gross impiety of Wolff by arguing that his doctrine of pre-established harmony entails, among other things, that soldier deserters cannot do otherwise. Finally on November 8, 1723, the King sacked Wolff as Professor of Mathematics at Halle (Wolff taught physics and philosophy, too) and gave Wolff a written order that Wolff would have 48 h to leave Prussia. Otherwise, Wolff would hang (Drechsler, 1997, 113–4).

Wolff left within 12 h and found refuge at the University of Marburg in the Landgraviate of Hesse-Kassel, where he became a celebrated teacher (ibid. and n. 27). Just when Frederick William I lifted the ban on Wolff's philosophy at Prussian universities in 1736 (Neumann, 2022, 148), the Pietist controversy was ignited again by a Wolffian translation of the *Five Books of Moses* ("Wertheim Bible") by an Enlightenment theologian Johann Lorenz Schmidt (1702–1749) (Goldenbaum, 2021, 7–8). The second wave of the controversy manifested in over 100 texts for and against Wolff (ibid.).

Lange was one of the main anti-Wolffian disputants in the second wave as well (ibid.). For instance, on May 10, 1736, he sent a document criticizing Wolff sharply to Frederick William I (Lachl, 1953, 571). The title of the document is "Kurzer Bericht derjenigen Lehrsätze, welche in der Wolffischen Philosophie der natürlichen und geoffenbarten Religion nachtheilig sind..." (*Brief account of those doctrines in Wolffian philosophy which are detrimental to natural and reveaked religion*). Wolff responded quickly with "Ausführliche Beantwortung der ungegründeten Anschuldigungen Herrn D. Langes..." (*Detailed reply to the unfounded allegations by Mr. D. Lange*). This exchange from May 1736 was immediately translated into French by the Prussian court (ibid.). It was published as *Recueil de nouvelles pieces philosophiques, concernant le different renouvellè entre messieurs Joachim Lange ... et Chretien Wolf ... avec des avis au lecteur, contenant l'histoire de ce different* (*Collection of new philosophical works concerning the renewed difference between misters Joachim Lange ... and Christian Wolff ... including readings containing the history of the said difference*).[16]

Lange's main criticism of Wolff's account is that it does not satisfy a necessary condition for a free action: that the will must make an efficacious causal contribution to the physical action independent from the intellect or the senses. For Lange then, we can will by the mere power of the will; Lange is a voluntarist rather than an intellectual compatibilist. In Wolff's view, the will is necessitated by the understanding and reason that provide the sufficient ground for the will by representing distinctly what is good. Furthermore, the will cannot have any efficacious causal contribution due to his rejection of physical influx between the soul and body. Therefore, Lange claims that Wolff reduces human agency to the working of a clock having an internal necessitating principle: the unrestricted PSR (Dyck forthcoming). This, as Lange puts it in a 1724 critique of Wolff's metaphysics, is "Stoicism" or "partial Spinozism"

[16] Its second edition from 1737 is at our disposal (Lange, 1737; Wolff, 1737).

leading to "Atheism": determinism (everything occurs necessarily) and "absolute fatalism" (Lange, 1724, 152–154, transl. in Dyck, 2019). Lange repeats this criticism more or less in his document from May 1736 and adding the censure of the Wertheim Bible (Lange, 1737).

9.5 Wolff in Sheep's Clothing: Was Du Châtelet Acquainted with Wolff on Free Will in 1737?

While the second wave of the Pietist debate was still raging in August of 1736, the son of Frederick William I, Crown Prince Frederick, began correspondence with Voltaire. At the time, Voltaire was hiding at Du Châtelet's Châteu de Cirey in North-Eastern France, in exile from the public outroar that had occurred from pubishing the *Letters to the English* at Rouen two years earlier (see Neumann, 2022, 168–172; cf. Prunea-Bretonnet, 2019; Vaillot, 1988, Chap. 1). Along with the first letter from August 8, Frederick sent Voltaire the French translation of the exchange between Lange (1737) and Wolff (1737) from May 1736. He likewise promised to send the French translation of Wolff's *German Metaphysics*, done by his friend, a Saxon diplomat Ulrich Friedrich of Suhm (1691–1740), once it was ready (D1126.[17]).

This was the beginning of a long series of correspondence between the two, which included Voltaire sending *On Freedom* to Frederick in October 1737 (St. P. ms.; D1376). Du Châtelet is largely present in the correspondence as well, and there is no doubt that she had access to, and probably read with great attention, everything that was exchanged between the two men (see e.g., D1251 and D1307; cf. D1265; Du Châtelet, 2018 I:281–282). This raises the obvious question of the influence of Wolff's *German Metaphysics* and the Pietist debate on the composition of *On Freedom.*

Since Suhm had not finished the translation by Fall, Frederick told Voltaire on November 4, 1736, that he will make a copy of what was done ("*ce qui est achevé*") and send it to Voltaire as an extract, promising the rest later (D1188). He finally fulfilled this promise by sending an extract of the *German Metaphysics* around December 1736 or January 1737, while Voltaire was visiting Amsterdam (D1247). We know from the correspondence between Suhm and Frederick that Suhm had managed to translate the *German Metaphysics* to the end of Chap. 3, which includes Wolff's discussion of free will, by June 16, 1736 (cf. Neumann, 2014, 52–54, n. 39; Preuss, 1846–1856, vol. 16, 270). Taking Frederick at his word, we have good reasons to suppose Wolff's account of free will to have been included already in this first extract.

Voltaire reportedly received the first part of *German Metaphysics* in January 1737 (D1225) and returned to Cirey in February (D1289). At the end of March, Voltaire

[17] References to the correspondence between Frederick and Voltaire are to the *Oeuvres complètes de Voltaire*, vol. 88 (Besterman 1969). We have followed the standard practice of indicating each letter with the code D 'order number'. Translations of the letters are by Forsman.

reported to Frederick that he and Du Châtelet had read the first extract of Wolff's *German Metaphysics* "with great attention" between February and March 1737, while working together closely at Cirey (D1307).

Frederick sent the second part of the translation, "until paragraph 770", on May 8[18] and said that the rest would arrive with his ambassador (and probable lover) Baron Dietrich von Keyserling (1698–1745), whom he sent to Cirey for a private visit that Summer (D1323; cf. D1330). Voltaire received the second part around the beginning of June (D1334) and he and Du Châtelet received Keyserling as their guest at Cirey in July 1737 (D1359). Note that Wolff discusses the pre-established harmony in §§753–67 of Chap. 5, and therefore the discussion should have been included with this second part of the translation.

Finally, around mid-October, Voltaire reports that he has "after a thousand detours" received "the third notebook [*cahier*]" of Wolffian metaphysics and comments and critiques Wolff's work as a whole (D1375). This mention of a "third notebook" appears rather surprising (typically the translation is mentioned to have been sent in two parts, e.g., Neumann, 2014, 2022; Prunea-Bretonnet, 2019). However, the simplest answer is that this was the last part of the *German Metaphysics*, from §771 onwards that Frederick said would arrive with Keyserling. For some reason, this seems to have not transpired and Voltaire instead received the third part three months later.[19] While this third part obviously would have arrived too late to have had a real impact on the writing or editing of *On Freedom*, Wolff's account of free will and pre-established harmony had already been included in the previous parts of the translation. Thus, our argument for the influence of Wolff's account on Du Châtelet is not compromised.

Going by the timeline of Voltaire taking the first part of the translation to Cirey in February, and sending *On Freedom* to Fredrick in late October, gives Du Châtelet about half a year to study Wolff on free will and action. Wolff discusses freedom close to the end of Chap. 3 (§§404–521) which is an exposition of empirical psychology (*On the Soul in General*). Wolff's discussion of freedom and the will is rather brief and clear; he is systematic in his typical style and provides definitions of central concepts. A couple of months is enough for understanding the basics of Wolff's theory, especially to such a sharp mind as Du Châtelet's. We find it therefore likely that Du Châtelet was familiar with Wolff's account of free will and action while she wrote or at least finalized *On Freedom*. Therefore, it is likewise probable that Wolff's account influenced Du Châtelet's philosophy of freedom in one way or another.

It is also likely that Du Châtelet knew about the Pietists controversy around this time. At the very end of Chap. 3, which Du Châtelet had access to since February 1737, Wolff insinuates the pre-established harmony. His proper discussion of it is

[18] Neumann (2022, 169) suggests Frederick to have sent the second part already in February. However, in the Letter from February 8, 1737, Frederick only states that he *hopes* ("*J 'espère*") to send the translation soon (D1281).

[19] One suggestion could be that this was the second part of the translation that Voltaire received in October (cf. Prunea-Bretonnet, 2019, 303). But this does not fit the evidence as Voltaire reports having the second part at the end of July (D1359) and he clearly indicates the Wolffian text in October as the "third notebook" (D1375).

in Chap. 5, in the exposition of rational psychology (*On the Essence of the Soul*), which Du Châtelet gained access to in around June 1737. Also recall that in his very first letter, Frederick sent the May 1736 exchange between Lange (1737) and Wolff (1737), where Lange recapitulates his criticism from 1724 to Voltaire in the French translation. While Lange does not discuss the pre-established harmony in detail in the work in question (he dismisses it in less than half a page, calling it ridiculous and referring to its consequences; (Lange, 1737, 46–47) Wolff explains it in length in his response to Lange (Wolff, 1737).

Voltaire and Du Châtelet therefore had detailed information of the Pietist debate, straight from the two central figures in the debate. Voltaire was also very interested in both the Pietist debate and the discussion of atheism from his first letter to Fredrick onward (c. September 1, 1737; D1139) and later commented on the pre-established harmony and the accusations targeted against it (c. January 15, 1737; D1255).[20] This indicates that at least Voltaire was both a) interested in and b) rather knowledgeable of the details of the debate. And while we have to speculate more on Du Châtelet's interest, we can at least reasonably expect her to have known as much of the details as Voltaire did because of their close co-operation at the time.[21]

Of course, it is possible that Du Châtelet had heard of Wolff's philosophy and the Pietist debate before Fredrick sent the translation; Wolff was already a famous philosopher by the time and his sacking from Halle due to the Pietist influence had been a large scandal, the influence which was still felt throughout Europe at the time.[22] However, we have found no evidence that Du Châtelet would have had close connection with Wolff's philosophy or the Pietist debate before early 1737. Du Châtelet began her correspondence with Fredrick in 1738 and with Wolff most likely around 1739, when she was already studying his philosophy more intensely under the tutoring of Johann Samuel König (1712–1757).[23] In light of the evidence known to us, reading the texts sent by Fredrick was the first time Du Châtelet got into close contact with Wolff and the Pietist debacle.

Having established the probable connection that Du Châtelet had to both Wolff's account of free will and the Pietist debate, we can now move on to the last part of the paper, where we discuss *On Freedom* in relation to this context. We approach the issue specifically with regards to (1) the connection between the will and physical

---

[20] This is the same letter in which Voltaire reports about receiving the first part of the GEM translation.

[21] Suhm's translation of GEM, which previously was thought to be lost (see Droysen 1909), was recently found in two parts by Hanns-Peter Neumann, with the first part (consisting paragraphs §§1–770, so the first two parts Fredrick sent) located in Bibliotheka Jagiellonska in Kraków, and the second part (consisting paragraphs §§771–1089, so the third part which Voltaire received in October 1737) located in Fredrick's former private library in Schloss Charlottenburg in Berlin (Neumann, 2022, 152).

[22] For a possible second connection on how Du Chàtelet might have been familiar with Wolff and the Pietists, see Hagengruber (2019, 2021).

[23] See Neumann (2022, 150). Du Châtelet mentions working with König in a letter to Fredrick (February 27, 1739) (Preuss, 1846–1856, vol. 17, 24; Du Chàtelet, 2018, I:500).

action and (2) whether Du Châtelet purports a compatibilist or libertarian view of freedom. Regarding the latter, we propose a conciliatory reading.

9.6 Du Châtelet on Freedom

Comparing Du Châtelet's *On Freedom* with Wolff's account of free will in the *German Metaphysics*, it is first of all important to make a distinction between the *freedom of the will* and the *freedom of (physical) action.* This distinction runs deep into the debate between compatibilist and libertarian accounts of freedom. Briefly, compatibilists, like Wolff, consider freedom to be a kind of self-determination and typically emphasize the freedom of the will over the freedom of physical action (or do not make a distinction between the two). By contrast, libertarians, like Clarke,[24] consider the power to do otherwise essential for freedom and typically emphasize the freedom of physical action over the freedom of the will. Of course, freedom of the will and freedom of action are not mutually exclusive; a compatibilist can similarly be worried about the freedom of action and a libertarian can consider the will to be free as well (free actions follow from free will). The question is more about where the emphasis lies rather than which one trumps the other.

When it comes to freedom of the *will*, Du Châtelet and Wolff agree on several key aspects. Both consider that external determination is incompatible with freedom (e.g., OF 494, 496), that the will and physical motion are distinct entities (e.g., OF 493), and that the will requires reasons or motivations to work as it should (e.g., OF 496). Also, Du Châtelet began to support the PSR at least by 1739–1740, as we know from early drafts of *The Foundations of Physics* from the period (see Janik, 1982; although, unlike Wolff, Du Châtelet restricts the PSR to mainly contingent truths, IP I.8; cf. Wells, 2021). We do not know for certain whether Du Châtelet already supported the PSR in 1737.[25] However, since she and Voltaire read the French translation of the *German Metaphysics*, which includes an unrestricted PSR in terms of actuality in the beginning of the second chapter, at Cirey in early 1737, she must have at least been aware of the philosophical concept in question.[26] Whether she already fully supported it at the time is, of course, impossible to say just based on her familiarity of the concept. But considering the influence that the *German Metaphysics* had on Du Châtelet's metaphysical thinking (see Neumann, 2022, 169–171), it is not out of

[24] There is debate whether Clarke is fully a libertarian or whether there are compatibilists elements in his account (e.g., Yenter & Vailati, 2021, Sec. 3.2). We will not take part in this issue and instead follow the standard interpretation of Clarke as a libertarian. See also Sangiacomo (2018), Jorati (2019, esp. 264. n. 29), Wells (2022).

[25] There are some early drafts of the *Foundations* from around the time but, unfortunately, it is difficult be sure if they already included Chap. 1 with the PSR, as chapters were worked on out of order (Paris ms.).

[26] Another source for her to know about the PSR is the Leibniz-Clarke correspondence, which she was familiar with by the late 1730's (Du Châtelet, 2018, I:367; cf. Hutton, 2012, 83–84; Wells, 2022, 224).

the question. If Du Châtelet supported the PSR already in 1737, it would bring her account of freedom even closer to Wolff's view. Every single free action would have to have a sufficient reason (whether internal or external) for that action to have taken place, without the determination of that reason being incompatible with the freedom of the action.[27]

However, it is possible to distinguish at least two main differences between Wolff's and Du Châtelet's accounts of freedom in the texts in question. The first one relates to how the will and its motivations are connected to our physical actions. The second one relates to locating the source of our freedom.

9.6.1 Will and Physical Action

We will start with the first. Du Châtelet gives only two explicit characterizations of the will in *On Freedom*, both very close to Wolff's definition. She considers the will (*volonté*) "the last perception or approval [*approbation*] of the understanding [*l'entendement*]" (OF 493) and especially notes that "one easily feels that willing, judging, etc. are merely different functions [*fonctions*] of our understanding" (OF 492).[28] These descriptions clarify that for Du Châtelet, the understanding and the will are not numerically distinct faculties or powers but rather a single power with different functions. (Cf. GEM [1719], §§492 and 745.) And like Wolff, Du Châtelet is committed to the internal determination of the will and its required motivations:

> Thus, due to his intelligence, man is under the necessity of willing that which his judgment presents to him as best. If it were otherwise, he would have to be subject to a determination by something outside of himself, and he would no longer be free; for to will that which does not please is a genuine contradiction; and being free means doing what one judges to be best, what brings pleasure. We can hardly conceive of a being that has more freedom than the capacity of doing what pleases him. (OF 496.)

In other words, our will (the last approval of the understanding) is internally determined by what our understanding judges to be the best. All this is very much in line with Wolff, as was seen above. Likewise, for Du Châtelet, the will cannot be the *efficient cause* of our physical action; however, it can be its *occasion*:

> [T] he will is never the cause of our actions, even though it is their occasion; for an abstract notion cannot have any physical influence over the physical self-moving power, which man

[27] There are some passages in OF which could be read as an indirect support of the PSR, for example: "If it were to follow that man is not free, because his will is always determined by the things that his understanding judges to be best, it would follow that God is not free, and that everything in the universe would be effect without cause, which is absurd" (OF 495).

[28] There are likewise obvious connections to Descartes's account of free will here, which Du Chàtelet was certainly aware of (see Du Châtelet, 2018, I:367). Note, however, that unlike Wolff, Descartes is not an intellectualist compatibilist but a *voluntarist* (see e.g., Forsman, 2017). And like in the case of Du Chatelet's OF, there is a lot of scholarly debate on whether Descartes is a compatibilist or a libertarian (see e.g., Newman, 2023).

possesses; and this power is exactly the same, before and after the last judgment of the understanding. (OF 493)

As mentioned, these changes in the will being the occasion of our bodily actions has caused perplexment and debate among scholars, with attempts to explain the issue by referring to Clarke's usage of *occasional causation* (Jorati, 2019). However, if Du Châtelet was influenced by Clarke here, it obviously raises the question whether her account of freedom would in fact be libertarian, like Clarke's, instead of compatibilist, like Wolff's account (Wells, 2022 argues for the libertarian reading).

We believe that contextualizing *On Freedom* with Wolff's account of free will and the following Pietist controversy may help in understanding this passage. As discussed, one of the main issues in the controversy was the relationship between our will and our physical action. Wolff famously considers the *pre-established harmony* to be the most likely hypothesis for the community between the soul and the body, rejecting the other options of *physical influx* (the soul and the body acting upon each other through an efficiently causal inflow of attributes between the two) and *Cartesian occasionalism* (the soul and the body acting upon each other through God's immediate action). According to Wolff's pre-established harmony, there is no natural influence between the two; they simply operate in unison with each other because God has pre-established a perfect harmony between them (GEM [1719], §760–765). Lange objected that Wolff's account denies any efficacious causal contribution from the will to the action, amounting to fatalism and atheism (Lange, 1724, 152–153; 1737, 34–40.)

Nonetheless, Lange's favored solution to the communication of the two substances, physical influx, is ruled out by Wolff as being against the conservation of motor force and thus against the laws of nature. Du Châtelet seemingly rejects it for the same reason:

[T[he will, that is, the last perception or approval of the understanding […] cannot have any influence over the self-moving power... […] [F]or an abstract notion cannot have any physical influence over the physical self-moving power... (OF 493.)[29]

There is nothing in *On Freedom* that explicitly indicates Du Châtelet's preference for pre-established harmony as the best explanation of the relation between the will and physical action. We can only speculate why this is so, especially since Wolffian pre-established harmony is restricted to the harmony of the soul and body, unlike Leibniz, whose pre-established harmony holds among *all* created monads (*Monadology*, §7 [AG, 213–214]; cf. Wells, 2022, 236, n. 7). However, in light of the Pietist critique, she may have avoided mentioning pre-established harmony due

[29] One could question this reading by noting that the will is not necessarily *an abstract notion* and therefore Du Châtelet does not explicitly reject physical influx in OF. While this is true, she does elsewhere worry that the conservation laws of motion might conflict with human freedom, indicating resistance to the influx (Letter to Maupertuis, April 30, 1738; Du Châtelet, 2018, I:342; see also Wells, 2022, 227 and n.10). Also, note how she phrases the issue: the will is simply the "*last perception or approval of understanding*", granting the reasons or motivations for the action to be good. Reasons or motivations are "abstract notions".

to its association with fatalism; an association that she would have wanted to avoid for obvious reasons.

If we were to suppose that Du Châtelet is influenced here by Clarke's account, we could read her as favoring the third option: occasionalism. But it is crucial to note that Du Châtelet's account here would not be what Wolff views as Cartesian occasionalism, by which he most likely means Malebranchian occasionalism, but instead *Clarke's occasionalism.* The former position is roughly that God is the only efficient or productive cause in the world; created things are merely occasions for the efficient causation that God produces.[30] The latter position allows also for created intelligent agents to be at least partially causally efficient on (non-efficient) bodies as well as on God's occasion (see Sangiacomo, 2018). Since Wolff's criticism of occasionalism is directed only at Cartesian occasionalism, it does not apply to occasionalism in this latter sense.[31]

Du Châtelet could have then supported Clarkean occasionalism as an alternate explanation to physical influx, Cartesian occasionalism, or pre-established harmony for the causal interaction between the will and physical action. This way, she could view the will having a true and genuine causal contribution to the physical action (which was the main Pietist critique of Wolff's account), without the will being completely independent from the intellect, and without giving up on the will's internal determination by reasons or motives.

However, note that the textual evidence for the occasionalist reading is very scarce. The explicit evidence consists of one word. Moreover, as was seen above, Wolff similarly uses occasion language in the *German Metaphysics* without considering it conflicting with pre-established harmony. Du Châtelet might simply mean that our will and our physical action are harmonious, without going into a detailed explanation why this is the case. The previous draft of *On Freedom*, in fact, begins with setting aside the question whether conservation of motion is contrary to our freedom (St. P. ms.; Barber, 1989, 484, n. 1). This can be seen as a further indication that Du Châtelet is not willing to make a detailed account even in the more developed version. Nevertheless, the fact that she does not clearly indicate even sympathy with pre-established harmony as the likely connection between the will and physical action distinguishes her account from Wolff's. The Pietist controversy context may help us make sense of this difference.

---

[30] There is debate whether this sort of occasionalism is actually all that different from pre-established harmony (e.g., Nadler, 1993). We do not take sides in this issue.

[31] Leibniz's criticism of occasionalism might have been so influential in Germany that other forms of it were not recognized, or at least not considered. Occasionalism remained, in various forms, a popular theory of causal explanation in France and England. See Dyck (forthcoming). One example of occasionalist theory of causation in English philosophy is Margaret Cavendish's (1623–1673) *occasional causation*, according to which self-moving bodies themselves are the efficient cause of their motions while the motion of other bodies is merely a *triggering or redirecting* occasion (see e.g., Cunning, 2022; Lascano, 2023). However, we are not aware of any evidence that Du Châtelet was acquainted with Cavendish's thought by the writing of OF. Apparently, only very few early moderns were acquainted with Cavendish's philosophy.

9.6.2 *Freedom of the Will and Freedom of Action*

Let us then move to the second distinction. Relating to the previous discussion between freedom of the will and freedom of physical action, Du Châtelet's characterization of freedom does not always seem to be in line. Sometimes she characterizes freedom in a similar way to Wolff as the freedom of the will that is determined by what the understanding judges to be best.

> I call freedom the *power of thinking* [*pouvoir de penser*] of one thing or of not thinking of it, of moving of not moving, in accordance with the *choice of one's own soul* (*choix de son propre esprit*). (OF 485; translation altered; emphasis added.)

> Is doing what we want [v*eut*] and what we *judge to be most advantageous* [...] not exactly the s*ame as being free?* (OF 495–496; emphasis added.)

> *To will that* which does not *please* [*plaisir*] is a *genuine* [*véritable*] *contradiction*; and being free means *doing what one judges to be best* [*le meilleur*], what brings pleasure. (OF 496; emphasis added.)

Compare these with Wolff's characterization: "*freedom* is nothing other than the *capacity of the soul* through its own spontaneity, for *choosing* that one among two equally possible things that *pleases* [*gefället*] *it most*" (GEM [1719], §519; the three last emphases added). The similarity is striking, especially with the discussion of the choice being one that pleases the soul.[32]

However, unlike Wolff, there are also places in *On Freedom* where Du Châtelet clearly emphasizes the *freedom of action* instead of the freedom of the will. For example, after noting that the will is never the cause of our actions but can be their occasion, she considers that "the physical power of acting is thus what makes man a free being, regardless of the use that he makes of it" (OF 494). Also, recall the Buridan's ass passage:

> [W]hat is quite certain is that the *self-moving power* [*pouvoir soi-mouvant*], the *sole and true* [*seule et véritable*] *source of freedom*, cannot be destroyed [*détruit*] by the indiscernibility of two objects; for insofar man has this self-moving power, man is free (OF 493).

Not only does the passage establish our power of self-motion, our ability to act physically, as the *true and actual* source of our freedom, but it likewise raises an interpretative question that cuts right to the core of the compatibilist debate on the text. If Du Châtelet means here that we can choose to perform a physical action even in a case of a perfect indiscernibility of two objects that please our soul to the exact same degree, if the ass can actually decide to move towards one of the perfectly similar haystacks in identical distance, then she would clearly go against Wolff's account, which states:

> Since without motives one can neither will or not will, it is *impossible that someone wills something or not* when motives of *equal weight* are present on each side. (GEM [1719], §508.)

[32] Unfortunately, we did not have access to Suhm's French translation of GEM at the time of writing and cannot therefore compare Du Châtelet's French with that of the GEM translation they read with Voltaire. But the similarity is striking even across German and French.

Wolff does not think that bodies have a self-moving power; they have only force to move other bodies and to resist motion, as was seen above. If read like this, the passage by Du Châtelet would be the clearest textual evidence for a libertarian reading in the text, and the supporters of a libertarian view of Du Châtelet have made a lot of use of it (see esp. Wells, 2022).

However, we do not think the libertarian reading of the Buridan's ass passage is all that given. Recall that for Wolff, even in the case of the strongest possible internal determination, it is always *absolutely possible* to do otherwise; it just is not *morally possible.* Du Châtelet makes a similar distinction in *On Freedom* almost immediately after the Buridan's ass passage:

> [O]ne must carefully distinguish between *physical necessity* [*nécessité physique*] and *moral necessity* [*nécessité morale*]. The former is always *absolute*, but the latter is always *contingent*, and this moral necessity is *entirely compatible* with the most perfect natural and physical freedom. (OF 494.)

This distinction allows for an alternate reading of the Buridan's ass passage. What Du Châtelet could mean here is simply that in a case of perfect indiscernibility of motives, it is still a*bsolutely possible* for us to move; it just is not morally possible. Taking the self-moving power to simply mean our bodily capability to perform physical actions, even a compatibilist could claim that this power is not destroyed by an equal weight of reasons; it just cannot be *exercised* in such an instant. This reading would also explain what Du Châtelet means with the self-moving power being "exactly the same, before and after the last judgment of the understanding" (OF 493). Whether we will or not will a certain action, our body retains the physical ability to move.

Of course, this alternate reading of the Buridan's ass passage does not conclusively settle the question for compatibilism, for it is still the case that Du Châtelet tends to locate freedom more in the self-moving power in certain places in the text, emphasizing freedom of action over freedom of the will. Therefore, she considers the *power of self-motion* to be relevant for freedom, in contrast to Wolff. One can indeed see a possible Clarkean influence here (following e.g., Jorati, 2019; Wells, 2022). However, this leaves her position in relation to compatibilism or libertarianism rather open in the text. Certain passages, where she emphasizes the will's determination for what it perceives as the best to be essential for freedom, seem to support a more compatibilist reading, while other passages, where she considers the self-moving power as the true source of freedom, seem to support a more libertarian reading. It is possible that she simply had not made up her mind over the issue when writing the text. Some of her phrasings would seem to support this view, as she writes in relation to the Buridan's ass passage that:

> With respect to the power of choosing between perfectly similar things, it is difficult to be able to say what would happen to us then, since we do not know that power. I do not even know if this power would be a perfection... (OF 493.)

Her ambivalence could be taken as a sign that she wants to retain the self-moving power's ability for movement even in the case of perfect equal balance of motives but does not wish to commit to a clear stand on whether it retains it by more compatibilist or libertarian means.[33] Just like in the case of the will's connection with physical activity, it appears that Du Châtelet is not willing to commit to a specific position, as her main interest is to reply to "the illustrious opponents" of freedom (OF 485).[34] Nevertheless, her emphasis on the essentiality of self-moving power for freedom clearly distinguishes her from Wolff. And just like in the case of will and physical activity, the context in Wolff's account of freedom and the following Pietist debate can illuminate this ambivalence.

9.7 Conclusion

In this article, we have contextualized Du Châtelet's *On Freedom* with the account of freedom and the will by Wolff, as he purports them in the *German Metaphysics*, and the following debate with the German Pietists, who considered Wolff's account leading to fatalism and atheism. In the second section, we described the history of *On Freedom* and problematized the current scholarly debate on its reading. In the third section, we briefly considered Wolff's account in the *German Metaphysics*, which Du Châtelet was familiar with when writing *On Freedom.* In the fourth section, we discussed the accusations by the Pietists, especially those of Joachim Lange, on Wolff's account of freedom and the connection between the will and physical action. In the fifth section, we considered and demonstrated textual and historical evidence for the influence of Wolff's account on Du Châtelet and for her knowledge of the Pietist debate. In the sixth section, we used this contextualization to clarify certain puzzling parts of *On Freedom,* particularly those having to do with the connection between the will and physical action and whether Du Châtelet supports a compatibilist or libertarian account of freedom in the text. Regarding the first, we suggest a reading where the context of the Pietist critique of Wolff's pre-established harmony helps us understand her view on the matter. Regarding the latter, we suggest a conciliatory reading due to textual evidence for both compatibilist and libertarian interpretations.

[33] See also the Letter to Maupertuis, April 30, 1738 (Du Châtelet, 2018, I:342), where she expresses uncertainty over the issue due to conservation laws. A later letter (May 9, 1738; Du Châtelet, 2018, I:347) indicates a clearer stand towards libertarian freedom, while still expressing uncertainty over the compatibility with conservation (Cf. Wells, 2022, 226–227). However, since this letter is of a later date, it can hardly be used for support on her account in OF, especially because Du Châtelet's thinking progressed so quickly in those times.

[34] Who might she be referring to here? Bayle is a possibility. Other strong candidates include Hobbes and certain Calvinistic thinkers. Wolff, however, does not seem to belong to her opponents (unlike Barber implies, 1989, 387), a sentiment where she might disagree with Voltaire who opposes Wolffianism in his discussions with Frederick on the topic (esp. D1432). In a later letter, from March 10, 1740, Voltaire writes to Frederick, in a tone that does not sound overly pleased, that Émilie has become "completely Leibnizian" (D2177).

References

Barber, W. H. (Ed.). (1989). *Traité de métaphysique*: Critical edition. In *Oeuvres complètes de Voltaire* (Vol. 14, pp. 357–503). Voltaire Foundation.

Besterman, T. (Ed.). (1969). *Oeuvres complètes de Voltaire* (Vol. 88). University of Toronto Press.

Cunning, D. (2022). Margaret Lucas Cavendish. In E. N. Zalta, & U. Nodelman, (Eds.), *The stanford encyclopedia of philosophy* (Winter 2022 Edition). https://plato.stanford.edu/archives/win2022/entries/margaret-cavendish/

Drechsler, W. (1997). Christian Wolff (1679–1754): A biographical essay. *European Journal of Law and Economics, 4*, 111–128.

Droysen, H. (1909). Die Marquise du Châtelet, Voltaire und der Philosoph Christian Wolff. *Zeitschrift Für Französische Sprache und Literatur, 35*, 226–248.

Du Châtelet, E. (c. 1737; 2020–2023; St. P. ms.) *Chap. 5. De la liberté*. The Saint Petersburg Manuscripts. A historical and critical online edition. Eds. R. E. Hagengruber, A. Brown, S. Ertz, & U. Kölving. https://historyofwomenphilosophers.org/stp/documents/view/chap.5.de_la_liberte

Du Châtelet, E. (c. 1737; 1989; OF). *Sur la liberté*. In *Barber 1989* (pp. 484–502).

Du Châtelet, E. (c. 1738–1940; 2021–2023; Paris ms.) *Institutiones de Physique,* The Paris Manuscript. Fr. 12265. A historical and critical online edition. Eds. R. E. Hagengruber, H.-P. Neumann, J. Muller, & A. Wells. https://historyofwomenphilosophers.org/dcpm/documents

Du Châtelet, E. (1740; IP) *Institutiones de Physique.* Prault.

Du Châtelet, E. (2009; IP) *The Foundations of Physics.* In J. Zinsser, & I. Bour, (Eds.). *Emilie Du Châtelet: Selected philosophical and scientific writings.* University of Chicago Press.

Du Châtelet, E. (2018). In U. Kölving, A. Brown, et al. (Eds.), *La Correspondence d'Émilie Du Châtelet* (Vol. I & II). Centre International d'Etude du XVIIIe Siècle.

Du Châtelet, E. (2021; OF) *On Freedom.* Translation by Jorati, J. *Project Vox.* Duke University Libraries.

Dyck, C. W. (Ed.). (2019). *Early Modern German Philosophy (1690–1750).* Oxford University Press.

Dyck, C. W. (forthcoming). Power, harmony, and freedom: Debating causation in 18th century Germany. In F. Beiser, & B. Look (Eds.). *Oxford handbook of eighteenth century German philosophy.* Oxford University Press.

Forsman, J. (2017). Descartes on will and suspension of judgment: Affectivity of the reasons for doubt. In G. Boros, J. Szalai, & O. I. Tóth (Eds.). *The concept of affectivity in early modern philosophy* (pp. 38–58). Eötvös University Press.

Goldenbaum, U. (2021). Who was afraid of Wolff's psychology? The historical context. In S. de Freitas Araujo, T. C. Ribeiro Pereira, & T. Sturm (Eds.), *The force of an Idea: New essays on Christian Wolff's psychology* (pp. 7–28). Springer.

Hagengruber, R. E. (2019). Relocating women in the history of philosophy and science: Emilie Du Châtelet (1706–1749), Laura Bassi (1711–1778), and Luise Gottsched (1713–1762) in Brucker's *Pinacotheca.* In S. Plastina, & E. M. de Tommaso (Eds.). *Filosofe e Scienziate in Etá Moderna* (pp. 123–136). Fabrizio Serra editore.

Hagengruber, R. E. (2021). "Mehr wert als tausend Männer" - Luise Aelgunde Victorie Kulmus Gottsched und die wahre Minerva. In J. Iffland, J. Imm, N. Jaeschke, & S. Schauberger (Eds.), *Musikwissenschaft der Vielfalt* (pp. 139–151). Allitera Verlag.

Hettche, M. & Dyck, C. W. (2019). Christian Wolff. In E. D. Zalta (Ed.), *The Stanford Encyclopedia of Philosophy* (Winter 2019 Edition). https://plato.stanford.edu/archives/win2019/entries/wolff-christian/.

Hutton, S. (2012). Between Newton and Leibniz: Emilie Du Châtelet and Samuel Clarke. In R. E. Hagengruber (Ed.). *Emilie Du Châtelet between Leibniz and Newton* (pp. 77–96). Springer.

Janik, L. G. (1982). Searching for the metaphysics of science: The structure and composition of madame Du Châtelet's *Institutions de physique*, 1737–1740. *Studies on Voltaire and the Eighteenth Century, 201*, 85–113.

Jorati, J. (2019). Du Châtelet on freedom, self-motion, and moral necessity. *Journal of the History of Philosophy, 57*(2), 255–280.

Lange, J. (1724; 2019). *A modest and detailed disclosure of the false and harmful philosophy in the Wolffian metaphysical system* (Bescheidene und Ausführiche Entdeckung der falschen und schädlichen Philosophie in dem Wolffianischen Systemate Metaphysico). In *Dyck 2019* (pp. 138–155).

Lange, J. (1737). Court expoise des aximes de la Philosophie de Mr. Wolf, qui sont prejudiciables a la religion naturelle, et a la revelation ou qui les detruit sent meme entierement l'unet lautre, en menant a l'atheisme par plusieurs detours, et sous de trompeuses apparences (Kurzer Bericht derjenigen Lehrsätze, welche in der Wolffischen Philosophie der natürlichen und geoffenbarten Religion nachtheilig sind, ja sie gar aufheben un gerawdesweges bei vieler gesuchter Vedeckung zum Atheismo verleiten). In *Recueil devouvelles pieces philosososophiques, concernant le different renouvellè entre messieuhrs Joachim Lange, Dr. et Professour en theologie a Halle, et Chretien Wolf, Professeur en philosophie a Marnpurg* (2nd Edn) (1st Edition in 1736).

Lascano, M. P. (2021). Émilie Du Châtelet on Illusions. *Journal of the American Philosophical Association, 7*(1), 1–19.

Lascano, M. P. (2023). *The metaphysics of Margaret Cavendish and Anne Conway: Monism, vitalism, and self-motion.* Oxford University Press.

Lascano, M. P. (forthcoming) "Du Châtelet on the Powers of Minds and Bodies" New Voices Talk Series, September 22, History of Women Philosophers and Scientists.

Lach, D. F. (1953). The Sinophilism of Christian Wolff (1679–1754). *Journal of the History of Ideas* 14(4), 561–574. https://doi.org/10.2307/2707702

Leibniz, G. W. F. (1714; 1989; AG). *Monadology* (*La Monadologie*). In R. Ariew, & D. Garber (Eds.), *G. W. Leibniz: Philosophical essays.* Hackett Publishing Company.

Nadler, S. (1993). Occasionalism and general will in Malebranche. *Journal of the History of Philosophy, 31*(1), 31–47.

Neumann, H.-P. (2014). Der preussische Kronprinz Friedrich und die französische Übersetzung der ›Deutschen Metaphysik‹ Christian Wolffs im Jahr 1736. *Forschungen Zur Brandenburgischen und Preussischen Geschichte, 24*(1), 35–68.

Neumann, H.-P. (2022). Émilie Du Châtelet within the correspondence between Christian Wolff and Ernst Christoph of Manteuffel. In R. E. Hagengruber (Ed.), *Époque Émilienne* (pp. 147–174). Springer.

Newman, L. (2023). Defense of a libertarian interpretation of descartes' account of judgment. *Pacific Philosophical Quarterly, 104*(3), 597–621.

Pattersson, H. T. (1938). Voltaire's "Traité de Metaphysique." *The Modern Language Review, 33*(2), 261–266.

Preuss, J. D. E. (ed.) (1846–1856). *Oeuvres de Frédéric le Grande* (Vols. 16 & 17). Imprimeur de Roi.

Prunea-Bretonnet, T. (2019). Émilie, Friedrich der Große und die „Leibniz-Wolff'sche" Metaphysik. In R. E. Hagengruber, & H. Hecht (Eds.) *Emilie du Châtelet Und Die Deutsche Aufklärung* (pp. 295–321). Springer Fachmedien Wiesbaden.

Sangiacomo, A. (2018). Samuel Clarke on agent causation, voluntarism, and occasionalism. *Science in Context, 31*(4), 421–456.

Vaillot, R. (1988). *Voltaire en son temps, vol. 2: Avec Madame Du Châtelet*, Voltaire Foundation.

Watkins, E. (Ed.). (2009). *Kant's critique of pure reason: Background source materials.* Cambridge University Press.

Wells, A. (2021). Du Châtelet on sufficient reason and empirical explanation. *Southern Journal of Philosophy, 59*(4), 629–655.

Wells, A. (2022). Du Châtelet's libertarianism. *History of Philosophy Quarterly, 38*(3), 219–241.

Wolff, C. (1719; 2019; GEM [1719]). *Rational thoughts concerning god, the world, and the human soul, and also all things in general.* In Dyck 2019 (pp. 99–134).

Wolff, C. (1732). *Psychologia empirica methodo scientifica pertractata.* Francofurti; Lipsiæ: prostat in Officina libraria Rengeriana.

Wolff, C. (1736; GEM [1736]) *Vernünftige Gedanken von Gott, der Welt und der Seele des Menschen, auch allen Dingen überhaupt, den Liebhabern der Wahrheit mitgetheilet, Die sechste Auflage hin und wieder vermehret* (1st edition 1719). Franckfurt und Leipzig.
Wolff, C. (1737). "Response de Mr. Chretien Wolf aux accusations mal fonde'es que mons. Lange (Ausführliche Beantwortung der ungegründeten Anschuldigungen Herrn D. Lange)". Translated by Jean Des Champs. *Recueil devouvelles pieces philososophiques, concernant le different renouvellè entre messieurs Joachim Lange, Dr. et Professour en theologie a Halle, et Chretien Wolf, Professeur en philosophie a Marnpurg.* 2nd Edition (1st Edition in 1736).
Yenter, T. & Vailati, E. (2021). Samuel Clarke. In *The stanford encyclopedia of philosophy* (Winter 2021 Edition). https://plato.stanford.edu/archives/win2021/entries/clarke/

Jan J. Forsman was Visiting Postdoctoral Scholar in Philosophy at the University of Iowa (2022–2024). His primary areas of research lie in Early Modern Philosophy and the History of Skepticism.

Jani I. Hakkarainen is Senior Lecturer at Tampere University. His primary research areas are Metaphysics and Hume.

Chapter 10
The Power of Love and the Force of Reason

Leibniz and Du Châtelet on the Foundations of Morality

Julia Borcherding

10.1 Introduction

While scholars of both Leibniz and Du Châtelet have made significant advances in addressing possible connections between them in regard to their theoretical philosophy—by discussing, for instance, Du Châtelet's use of Leibniz's two "great principles", the Principle of Non-Contradiction (hereafter: PC) and the Principle of Sufficient Reason (hereafter: PSR), or by investigating whether and to what extent Du Châtelet can be seen to offer a synthesis between Leibnizian metaphysics and Newtonian physics[1]—so far, little has been said about any possible points of connection between du Châtelet's and Leibniz's moral thought.[2] However, as I aim to show in what follows, here also, an intriguing set of continuities and divergences emerges.

I begin by showing, in Sect. 10.2, that for Leibniz as for Du Châtelet, a virtuous action is an action that must satisfy what we might call the moral PSR. Moreover, for

[1] Brading (2019, 1–25) provides an excellent overview of the scholarly discussion so far. Earlier commentators have tended to view the *Foundations* as trying to effect a "marriage between Leibnizian metaphysics and Newtonian science" (Barber 1967, 220), or as a merely synthetic attempt to place Newtonian physics on Leibnizian foundations [see e.g. Iltis (1977), Janik (1982)]. Newer readings tend to place greater emphasis on the transformative and original nature of Du Châtelet's work [see e.g. Brading (2019), Janiak (2018), Carus (2024a) and Carus (2024b)]. For Du Châtelet's treatment of the PSR and the PC, see Wells (2023, 2021), Blank (2019), Reichenberger (2016), Detlefsen (2019), Moriarty (2006), and Carus in the present volume.

[2] Compared to other areas of her thought, Du Châtelet's moral thought has been explored relatively little so far. For some notable exceptions, most of them focused on Du Châtelet's *Discourse on Happiness*, see Lascano (2023), Le Ru (2022), Blank (2022), Hagengruber (2017, 2011) and Rodrigues (2011).

J. Borcherding (✉)
University of Cambridge, Cambridge, UK
e-mail: jasb5@cam.ac.uk

C. Carus and J. K. McDonough (eds.), *Émilie Du Châtelet in Relation to Leibniz and Wolff*, Women in the History of Philosophy and Sciences 23,
https://doi.org/10.1007/978-3-031-80541-7_10

both, this commitment is deeply grounded in their anti-voluntarism. Furthermore, as I argue in Sect. 10.3, Leibniz and Du Châtelet also share a commitment to hedonism, though for neither philosopher does this commitment bottom out in an egoist ethics that invites us to simply pursue our own pleasure. Rather, both effect a reconciliation of self- and other-regarding motives by according a central role to the notion of love. Finally, both are committed to a version of moral nativism. Yet despite these convergences, I suggest in Sect. 10.4, Du Châtelet is ultimately perhaps best cast as effecting a sentimentalist transformation of Leibniz's rationalism when it comes to the moral realm. While she holds on to, and even amplifies, a commitment to certain crucial Leibnizian rationalist principles—importantly, the PSR—our passions and desires, though moulded and constrained by reason, now play a leading role in her own anatomy of morality.

10.2 Anti-Voluntarism and the Moral PSR

As this section will show, both Leibniz and Du Châtelet share a deep commitment to anti-voluntarism. As the positive flipside of this critique, both ascribe a central role to the PSR as a practical principle which provides a standard not only for theoretical reason, but also for moral action. I will begin by briefly sketching Leibniz's criticisms of voluntarism and his views on the moral role of the PSR, before moving on to Du Châtelet, who not only shares Leibniz's anti-voluntarist inclinations, but also much more explicitly appeals to the PSR as the foundational principle of moral action.

10.2.1 Leibniz

"To say *stat pro ratione voluntas*, my will takes the place of reason, is properly the motto of a tyrant", Leibniz writes in response to the Hobbesian voluntarist.[3] If God is to be no such tyrant, the moral law needs to be founded on God's reason, not on God's will. The divine will cannot command arbitrarily; rather, it needs to be determined by sufficient reasons that are grounded in the intrinsic goodness and perfection of the things themselves. For if God merely imposed his will through his absolute power (*de potentia absoluta*), then his rule would not be that of the most just, but only that of the most powerful being. He would then no longer be worthy of our love—a disastrous consequence for morality.[4] Hence, Leibniz concludes, when God chooses

[3] R 4. Please see the reference section at the end of this section for a list of all abbrevations used.

[4] R 71–2. See also G IV, 429/AG 37: "[T]hat God acts always in the most perfect and desirable way possible, is, in my judgment, the foundation of the love that we owe God in all things." On Leibniz's use of the distinction between *potentia Dei absoluta* and *potentia Dei ordinata*, see Backus (2014). As Backus points out, Leibniz holds that God possesses absolute power with respect to the realm of hypothetical possibles, but not with respect to the realm "of truth and being" which includes the eternal truths of both logic and morality.

to create this world, his will must be guided by his knowledge of the good. He knows which out of all the sets of compossible essences of things will yield the best world, and his choice, far from being an arbitrary one, must be guided by this knowledge.[5] While this might be thought to suggest that God's choice is thus necessitated by his omniscience and omnibenevolence, Leibniz insists that it was nonetheless a free one, if only we assume a sound understanding of freedom, not a libertarian freedom of indifference.[6]

This rejection of divine voluntarism is a central element of Leibnizian morality, and it applies to divine and human action alike. When we, as humans, act according to reasons grounded in what our intellect recognizes as the good, we mirror God by acting in accordance with the same motive of perfection that also guides divine action. Leibniz here makes it amply clear that the mere imitation or the mere following of the command of an authority is not enough to render an action just or virtuous. One must grasp the reasons that make the action a good one, and act in accordance with them. To act morally, therefore, is to act "according to reason"—to act guided by a value judgment which inclines the will towards this perceived good. The effort to act in accordance with this judgment, he asserts in the *Theodicy*, "forms the essence of will."[7] We aim to preserve and augment the good in ourselves, in others, in the world that surrounds us, just as God does in creation, where the values of the alternative possible worlds that present themselves to the divine intellect are the reasons which guide his will, and incline in him to choose the best. Freedom, in turn, is never freedom of indifference, but consists in our ability to act with reason in pursuit of the good. And insofar we do so, we imitate divine action:

> Regarding the image of God, I believe that our freedom represents it, not insofar as we are indifferent (since indifference is an imperfection), but insofar as we act according to reason (*entant nous agissons suivant la raison*).[8]

Freedom of indifference, Leibniz emphasizes, is no freedom at all, but imperfection. Like the divine will, our will is thus subject to the PSR. If we did not pursue what our intellect has determined to be good and "act according to reason", we would cease to be rational beings, and cease to be images of God. "[W]e do not will because we will to will", Leibniz writes in a short text titled *On Free Will*, "but because our nature is to will what we believe best. And this belief does not come from our will,

---

[5] Leibniz provides different descriptions of what the bestness of this world consists in, which has given rise to some scholarly debate. While interpreters such as Rescher (2003) argue that the best world is marked by an optimal balance between the variety of its phenomena and the simplicity of its laws, commentators such as Rutherford (1995, 394) suggest that the best world is the world with the maximal amount of perfection, where perfection is, first and foremost, understood as quantity of reality or essence, which in turn offers a maximal degree of intelligibility or "contemplatibility" for created minds, thus maximizing their happiness (Ibid., 402).

[6] According to the account given in the *Theodicy* (§288/H 303) a free action involves (i) intelligence, "which involves a clear knowledge of the object of deliberation" (ii) spontaneity, "whereby we determine", and (iii) contingency, i.e., "the exclusion of logical or metaphysical necessity".

[7] *Theodicy* §311/H 314.

[8] A II.2, 382.

but from the nature of things, or the state of our mind."[9] To believe otherwise, he adds, is to "subvert all the principles of morality."[10]

In later works, Leibniz often uses the concept of "moral necessity" to describe the rational determination that is central to moral action.[11] In the *Theodicy,* he writes, for example, "that God, that the sage, is carried to the best by a moral necessity."[12] If an action is morally necessary, it is obligatory, and an ideal moral agent—like God or the virtuous sage—cannot fail to perform it, even though the action is still free. Such an agent is guided by the "charity of the wise" (*caritas sapientis*), the other-regarding love which, together with the knowledge of the true principles of justice, characterises the virtuous Leibnizian moral agent. They know the best course of action available to them, and are compelled to this choice by a "happy necessity."[13] Moral necessity, Leibniz explains, is called "moral"

> because for the wise what is necessary and what is owing are equivalent things; and when it is always followed by its effect, as it is indeed in the perfectly wise, that is, in God, one can say that it is a happy necessity. The more nearly creatures approach this, the closer do they come to perfect felicity.[14]

While arbitrariness, therefore, is always contrary to justice, the more our will is determined by reasons provided by our knowledge of the good, the closer we approach this divine "happy necessity" in our own just actions, and the happiness which according to Leibniz results from them.

Next to regarding the PSR as metaphysical principle, which tells us "that we can find no true or existent fact, no true assertion, without there being a sufficient reason why it is thus and not otherwise", and as a corresponding constraint on our theoretical reasoning, Leibniz[15] thus also accords a regulative role to the PSR when it comes to the moral realm. Even though he never explicitly invokes the principle in this context, the PSR clearly serves as a principle of divine wisdom which compels

---

[9] A VI.4, 1408.

[10] A VI.3, 135.

[11] For this periodization and further discussion, see Grua (1953), 234–45, and Adams (2005). As Grua (1953), 222–26, points out, "moral necessity" was an expression extensively used in the jurisprudence of Leibniz's time.

[12] *Theodicy* §132/H 203 (transl. mod.).

[13] See, e.g., G.VI, 390/H 395: "Mr. Hobbes refuses to listen to anything about a moral necessity either, on the ground that everything really happens through physical causes. But one is nevertheless justified in making a great difference between the necessity which constrains the wise to do good, and which is termed moral, existing even in relation to God, and that blind necessity whereby according to Epicurus, Strato, Spinoza, and perhaps Mr. Hobbes, things exist without intelligence and without choice, and consequently without God. ...Moral necessity contains an obligation imposed by reason, which is always followed by its effect in the wise"; *Preliminary Dissertation* §2/H 74:"This fitness of things has also its rules and reasons, but it is the free choice of God, and not a geometrical necessity, which causes preference for what is fitting and brings it into existence. Thus one may say that physical necessity is founded on moral necessity, that is, on the wise one's choice which is worthy of his wisdom." Cf. *Theodicy* §§230, 234, 237, 282.

[14] H 387.

[15] G VI, 61/L 646.

God to create a world with an order harmonious enough to ensure the happiness of intelligent beings. Human morality, in turn, as a mirror image of the divine, equally seems to mandate the PSR as a foundational principle of practical reason. Finally, Leibniz at times emphasizes that the PSR is an epistemic first principle not only of physics, but also of morality, insofar as no demonstration of any theorem—physical or moral—could ever get off the ground without assuming it.[16] In this context, he also seems to hint at the idea—more clearly spelled out in Du Châtelet, as we shall see shortly—that the very existence of both science and morality depend on the PSR, because the assumption of the existence of a sufficient reason grounds our practices of enquiring further into the causes of a given phenomenon or the reasons which underlie a reason given, until we reach their ultimate explanations.[17]

When we turn our attention to Du Châtelet, we find both a very similar anti-voluntarist strand and an even more explicit embrace of the PSR as a governing principle of human morality. She explicitly identifies the PSR as a principle which not only operates in the theoretical realm, but also regulates our choice of moral norms and lets us adjudicate between them.

10.2.2 Du Châtelet

In her essay *On Freedom*, Du Châtelet, like Leibniz, argues that freedom of indifference—to act in the absence of reasons—is not real freedom at all, but merely a "freedom of idiots."[18] Moreover, she further claims, again like Leibniz, that the highest form of freedom is to be determined by "right reason", by what we perceive to be best, since "being determined by what appears to us to be the best, is an equally great perfection as the power to do what we have so judged."[19] This necessity to always do the best, she further explains in Leibnizian terms, is "only a moral necessity", not an "absolute necessity."[20] Finally, she likewise declares that the more "our determinations are founded on right reasons, the more we approach and imitate God's perfection."[21]

Du Châtelet's Leibnizian anti-voluntarism also comes out clearly in the *Foundations of Physics*, as well as in the scathing critique of the biblical idea of God in

16 G VII, 301.

17 A VI.3, 118.

18 SL 493. In one of the marginal comments Du Châtelet added to the *Traité de Metaphysique* [see Brown & Kölving (2003), 93], she equally criticizes the view that God could not be free if determined by a sufficient reason, arguing that God still acts freely if the reason in question originates in the divine intellect.

19 SL 494.

20 Ibid., 496.

21 Ibid., 494–5. For further discussions of Du Châtelet's conception of freedom, see Jorati (2019) and Wells (2022).

her *Examinations of the Bible*.[22] God, she argues there, cannot violate the laws of nature without causing a contradiction in his will. But he also cannot act without a reason, for, as Du Châtelet puts it in the *Foundations*, "supreme intelligence will not behave without intelligence."[23] Du Châtelet emphatically criticizes the God of the Old Testament, whom she casts as a wilful tyrant as cruel as Nero in pursuit of a self-serving, retributive model of justice, his only rationale the unfettered exercise of his will.[24] She complains that of the two twins, God chooses Jacob and reproves Esau without any sufficient reason, and she rejects the idea of divine grace that is received gratuitously.[25] Like Leibniz before her, Du Châtelet in the *Examinations* thus seems to suggest that a God who can be loved by us, who is truly worthy of his perfections, must act in accordance with a morality that can be judged by the natural light of reason and its foundational principles, the PC and the PSR.

In the *Foundations*, Du Châtelet similarly emphasizes that our scientific explanations ought not come to a halt through a simple appeal to the arbitrary will of God, as such an appeal would fail to satisfy the PSR, which demands an explanation not only for the actuality, but also for the possibility of the thing under consideration.[26] As for Leibniz, this demand is ultimately grounded in the idea that creation depends "on the will of God acting in accordance with his intelligence and wisdom."[27] But while Leibniz's anti-voluntarist views certainly play a part in grounding his ethical claims, he never explicitly introduces the PSR as a foundational principle of moral action. This step is now taken by Du Châtelet, when she, in Sect. 1.11 of the *Foundations*, not only declares the PSR one of the pillars of scientific knowledge, but also, as the section heading asserts, "the foundation of morality":

> *The Principle of Sufficient Reason is the Foundation of Morality.* The principle of sufficient reason is also the foundation of rules and customs that are only founded on what is called propriety (*convenance*), for since the same men may follow different customs, they can determine their actions in many ways; and when one chooses to prefer those to others for which there is more reason (*celles où il y a le plus de raison*), the action becomes good and could not be condemned; but it is said to be unreasonable when there are sufficient reasons for not committing it, and it is on these same principles that one custom can be judged better than another, that is to say, when it has more reason on its side.[28]

---

[22] Du Châtelet (1740), hereafter cited as IP; Du Châtelet (1742/2011), hereafter cited as EB. A partial translation of both works can be found in Zinsser and Bour (2009); hereafter cited as ZB. For further discussion of the *Examinations*, see the meticulous introduction to Schwarzbach's edition, as well as Schwarzbach (2022, 2008, 2006, 1995, 1993), Rey (2022), Seguin (2022, 2021), Hagengruber (2022), Krief (2020) and Le Ru (2019).

[23] IP 2.26/ZB 143.

[24] EB I.39; II.535.

[25] EB I.271; EB II.197, 401, 471, 480, 486.

[26] "Thus, one must admit nothing as true in Philosophy, when one can give no other reason of the possibility except for God's will, for this will does not make one understand how a thing is possible," IP 3.49.

[27] IP 3.26.

[28] IP 1.11/ZB 132 (translation modified).

How exactly ought we understand Du Châtelet's claim that the PSR not only regulates the formation of scientific hypotheses, but also governs moral action? In a recent paper, Marcy Lascano provides the following gloss on this passage:

> Du Châtelet claims that the principle of sufficient reason is the foundation of moral philosophy as well as of physic and metaphysics ... she argues that while different men may follow different customs, when we decide which custom has the most 'reason' behind it, we then say that this custom is good and right. Thus, the principle of sufficient reason provides a guide to action that must be used in order to achieve happiness.[29]

While this strikes me as a plausible summary of Du Châtelet's view, it still seems to leave open some important questions: What does it mean to say that a custom has "most reason" behind it? And do we just *say* that this custom is "good and right"? Or is there a deeper sense in which it actually is?

To begin with, Du Châtelet already observes in an earlier section of the *Foundations* that the PSR is a basic principle not only of theoretical, but also of practical reason insofar as "all humans naturally follow" it. No one, she claims against the voluntarist, "decides to do one thing rather than another without a sufficient reason that shows that this thing is preferable to the other."[30] At the same time, we are free to adopt different reasons for action, and thus can, as Du Châtelet puts it in the passage quoted above, "determine [our] actions in many ways." This, in turn, is a formulation that maps closely onto Du Châtelet description of contingent states of affairs earlier on in the *Foundations* as states that can be determined "in many ways".[31] As contingent actions, therefore, our free actions fall under the scope of the PSR as the principle of contingent truths.[32]

However, she further adds a normative dimension to this descriptive claim by arguing that the PSR can also serve a regulative function on the practical side, analogous to the regulative function she ascribes to the PSR in the realm of theoretical reason.[33] In the theoretical case, the PSR governs the formation and evaluation of scientific hypotheses, demanding not only that our hypotheses point to the possible causes of an observed phenomenon, but also that they provide us with reasons that can lead an intelligent being to understand not only *that* this phenomenon has occurred,

---

[29] Lascano (2021), 5.

[30] IP 1.8/ZB 128. Analogously, DC pronounces epistemic voluntarism to be psychologically impossible: "[W]e feel that we cannot force our mind to accept something without a sufficient reason," (Ibid., 129).

[31] IP 1.7/ZB 128.

[32] IP 1.8/ZB 128. For the same demarcation between the PSR as the principle of contingent and the PC as the principle of necessary truths, see e.g. G VII, 355–56.

[33] While there does seem to be a factive use of the PSR in Du Châtelet (as just pointed out, she appears, for instance, to regard it as a psychological fact about rational beings like us that we cannot act without a reason determining our action), she first and foremost introduces the PSR and the PC not as factive, but as regulative principles, which specify the conditions under which something can be intelligible to us in the first place. See Melamed and Lin (2023) for a helpful discussion of this distinction and its application to Du Châtelet.

but also *why* it does.[34] Applied to the moral realm, the PSR likewise demands a sufficient reason for our actions in this sense. When we want to hold someone responsible for their action, Du Châtelet explains, we persist in our query until we obtain a reason that explains why they acted so and not otherwise, one which fully satisfies our demand for a sufficient explanation.[35] Note that the PSR here not only seems to call for an explanatory, but for a *justificatory* reason. What we are after when we want to hold someone accountable is not simply a reason that explains why they have performed a certain action—say, because of a particular desire that has led them to pursue it—but a *good* reason, a reason, that is, which justifies their acting in in this way. The PSR here not only figures as a descriptive principle which states that we are, as a matter of psychological fact, compelled to act for reasons, as Du Châtelet argues in Sect. 1.8. Rather, as Sect. 1.11 makes clear, the PSR is also meant to serve as a normative principle which can help us determine the moral status of our individual actions, and of the rules or norms that may govern them. Our actions are good, Du Châtelet asserts, if we choose those for which we have sufficient justification and reject those which lack it. Similarly, while there are many possible moral rules which may differ from society to society and which "may determine our actions in many ways", one "can be judged better than another … when it has more reason on its side."[36] The PSR thus not only figures as a constraint on practical rationality, but also as a regulative principle governing our moral choices and norms.[37]

So does Du Châtelet here sharpen the moral rationalism we find in Leibniz? Or does she in fact take it into a different direction? Much, it seems, depends on Du Châtelet's answer to the question of what would count as a sufficient reason in a given instance, a question that the passage seems to leave very much open to interpretation. On the one hand, Du Châtelet's claim that we ought to choose those norms and actions which have "more reason on [their] side" and reject those for which we do not have sufficient reason seems to echo her Leibnizian claim in *On Freedom*, quoted earlier, that "the more our determinations are founded on *right* reasons, the more we approach and imitate God's perfection."[38] We might thus imagine, along Leibnizian lines, that any action that preserves or increases the perfection of a thing

---

[34] "Now a thing cannot come to exist without a sufficient reason, by which an intelligent being might understand why this thing becomes actual, having been possible before. … For any man who makes use of his reason must not be content with knowing that a thing is possible and that it exists, but he must also know the reason why it exists," IP 1.9/ZB 131. For further discussion of this requirement, see Wells (2021).

[35] IP 1.8/ZB 128.

[36] IP 1.11/ZB 132. For further discussion of this passage, see Borcherding (forthcoming).

[37] In this function, the PSR thus does not pose an explanatory demand that is a direct result of the world's actual metaphysical constitution (i.e., given that world is constituted in such and such a way, there is no effect that lacks a cause, and thus, there must be a reason for everything). Rather, it poses a normative demand, both on our theoretical explanations (where it rejects, e.g., any hypothesis that offers up an "explanation" for an event which does not actually advance our understanding (such as Molière's doctor's infamous dormitive virtue) and on our moral justifications (where it rejects, roughly, any maxim for action that does not advance our good, as we shall see in more detail below). Hagengruber (2017, 199) ascribes an analogous function to the PC.

[38] SL 494–5.

or the harmony of the whole may satisfy the moral PSR. On the other hand, however, Du Châtelet at the very start of the passage also glosses her claim that the PSR is the foundation of morality as the claim that it is "the foundation of rules and customs that are *only* founded on what is called propriety (*convenance*)."[39] This, it seems, could also suggest a relativist claim. As Du Châtelet goes on to say, "men may follow different customs", and so perhaps the reasons by which we are meant to judge the moral norms in question could be just as relative to a given society or culture as those norms themselves. Thus understood, her claims about the moral role of the PSR would certainly put some distance between Leibniz's moral views and her own.

10.3 Morality and the Pursuit of Happiness

Since the *Foundations* do not further expand on the moral role of the PSR and also do not offer any other direct evidence of Du Châtelet's views on ethics, we need to look elsewhere in order to determine whether these views align with Leibniz's, or whether they indeed constitute a significant departure. In the following sections, I will thus examine Du Châtelet's moral thought more broadly as we can reconstruct it from other writings, with a view to whether she parallels or diverges from views we find in Leibniz. I will then return to a consideration of Du Châtelet's moral PSR and how it should be understood.

10.3.1 Pleasure, Perfection and the Pursuit of Happiness

To begin with, Du Châtelet's writings reveal a deep commitment to psychological hedonism. All motivation for action, both human and divine, fundamentally derives from a basic striving for one's pleasure or happiness. "Willing what does not cause pleasure is a true contradiction," Du Châtelet writes in *On Freedom*, and in her *Discourse on Happiness*, she argues that the only true source of happiness are our "satisfied tastes and passions."[40] Indeed, she emphasizes, "[t]he only point of living is to experience agreeable sensations and feelings; and the stronger the agreeable feelings are, the happier one is."[41] Likewise, the satisfaction God derives from the best possible world serves him as the sufficient reason for creating it.[42]

[39] IP 1.11/ZB 132.

[40] SL 496; DH 350.

[41] DH 350. See also DH 365: "Thank goodness, it is up to us to choose the time of our death, if it is too slow in coming; but as long as we prefer to endure life, we must open ourselves to pleasure by all the doors leading to our soul; we have no other business." For further discussion of Du Châtelet's hedonism, see Lascano (2023), who argues that Du Châtelet does not defend a simple desire-satisfaction view, but a unique account of the passions as the motors of our happiness, where passions are not opposed to reason but often are what it is rational to pursue.

[42] IP 1.26/ZB 143.

We find a similar commitment to psychological hedonism in Leibniz. Like Du Châtelet, Leibniz identifies the pursuit of pleasure as the most fundamental motor of human action.[43] Moreover, for both Leibniz and Du Châtelet, this hedonism *prima facie* seems to come with an egoist streak: every deliberate action is directed towards our own good—towards that which we perceive to contribute to our own happiness—such that my pleasure or happiness always constitutes a sufficient reason for action.[44] This commitment in turn has led some interpreters of Leibniz to conclude that Leibniz is as uncompromising as Spinoza or Hobbes in holding that, as Nicholas Jolley puts it, "no one deliberately does anything except for the sake of his own welfare, for one seeks the good even of those whom we love for the sake of the pleasure we derive from their happiness."[45] However, neither Du Châtelet nor Leibniz in fact simply stop there, for both closely tie pleasure to perfection.

For any intelligent being, divine or human, Du Châtelet argues in the *Foundations*, "[t]he contemplation of perfection is the source of pleasure … for what has the most perfection pleases more, and a reasonable being only desires things in proportion to the perfections he notes in them." Both God and finite rational beings like us desire pleasure, and for each, the pleasure attained is proportional to the perfection perceived. But whereas divine knowledge is infinite, our understanding is limited, and so "we often mistake an apparent perfection for a real perfection", while God cannot be so deceived.[46] Mary Terrall has suggested that Du Châtelet's claims here in fact mark an important point of departure from Leibniz, insofar as she "reformulat[es] Leibniz's theology, in which God's choice of the most rational and efficient set of possibilities determines the actual world" to stress "the pleasure God takes in his creation."[47] However, as the following passage from Leibniz's *Meditation on the Common Concept of Justice* shows, there is indeed no such contrast:

> One cannot envisage in God any other motive than that of perfection, or, if you like, of his pleasure; supposing (according to my definition), that pleasure is nothing but a feeling of perfection, he has nothing to consider outside of himself; on the contrary, everything depends on him. But his goodness would not be supreme, if he did not aim at the good and at perfection as far as possible. But what will one say, if I show that the same motive has a place in truly virtuous and generous men, who supreme function is to imitate divinity, insofar as human nature is capable of it?[48]

---

[43] "For everything pleasant is sought for its own sake, and whatever is sought for its own sake is pleasant; all other things are sought for the sake of a pleasant thing, to the extent that they produce it, or preserve it, or destroy contrary things. All people sense this, whatever they may say; or at least they act according to it, whatever they may believe." (A VI.1, 464/L 136). See also NE 1, 2, 3/RB 90, where Leibniz prominently ascribes to both Locke and himself a version of psychological hedonism, according to which it is part of our very nature to strive for our own pleasure and to avoid pain.

[44] A VI.1, 41.

[45] Jolley (2005), 180. See also Hostler (1975), 47, and Brown (2011). Frey (2016), by contrast, argues that Leibniz was not an egoist or hedonist "in any non-trivial sense" (Ibid, 601).

[46] IP 1.26/ZB 143.

[47] Terrall (1995), 300–301.

[48] R 56–57.

Just like Du Châtelet, Leibniz here points to perfection as the motive of action that governs both the divine and the human will. Moreover, a closer investigation of the Leibnizian views which underwrite this claim can help us to understand more clearly the picture underlying Du Châtelet's own claims regarding the connection between pleasure and perfection.

First, Leibniz, in the passage just quoted and elsewhere, provides an explicit analysis of pleasure as a feeling or perception of perfection.[49] To Wolff, he explains that while we are unable "to give a nominal definition of pleasure, and pleasantness is no more familiar than pleasure",

> pleasure admits of a real definition, and I think it is nothing other than a sense of perfection. It is the same with regard to other clear but confused ideas; so a definition of the colour green can be given, but it is not nominal, but nevertheless it is real, containing the cause, as it is evidently constructed from blue and yellow."[50]

In the *New Essays*, Leibniz equally agrees with Locke that we cannot give a definition of pleasure that would enlist the marks by which it could be recognized, no more than we can give such "nominal" definitions of sensible qualities such as light, heat or colour. At the same time, we *can* give causal definitions of them—definitions which reveal the structure or inner constitution of their causes.[51] Sensible qualities such as light or heat can be reduced to minute mechanical qualities of bodies, and in perceiving such qualities "we certainly have no perception other than that of shapes and motions", albeit we only perceive them confusedly.[52] Likewise, in experiencing pleasure, we always also—albeit often confusedly—perceive its cause, namely the perfection of whatever has caused us to have this experience. While the notion of perfection at work here certainly needs to be understood broadly since our experiences of pleasure may be manifold, there is, Leibniz wants to hold, an underlying sense of perfection which nonetheless unifies them.[53] The perception of perfection, he argues, always fundamentally involves either a grasp of "harmony or order" or of its creation, where "harmony or order" is to be understood as "a state where everything is worthy of being observed, that is, the state of agreement or identity in variety; you can even say that it is the degree of contemplatibility."[54]

Second, as we already briefly saw above, Leibniz emphasizes that this pursuit of perfection not only constitutes the primary motive of divine action, but also the primary motive of a virtuous human agent, who—in so acting—performs an imitation of the divine. When we act truly virtuously, our choices are determined by our distinct grasp of perfection and we are moved by the happiness this perfection causes. Moreover, the greatest happiness we can attain, we derive from the perfection of most

---

[49] This analysis recurs frequently in Leibniz's writings. See e.g. RB 194; *Theodicy* §33/G VI, 122.

[50] GLW 18.

[51] RB 294. In the *Meditation on Truth, Knowledge, and Ideas*, Leibniz further adds that we can give a causal definition of a thing "if we understand the method by which the thing can be produced," L 293.

[52] A VI.4, 592: AG 27.

[53] A VI.4, 2794.

[54] G VI, 630/AG 233–34.

perfect being. God's perfections are infinite, and the pleasure that results from loving God is thus the greatest and most stable we can experience. And since we cannot love God without "willing what He wills", it follows, Leibniz argues, that "the greatest happiness, which causes one to love him, causes one to be happy and virtuous at the same time."[55]

Our innate tendency to pursue what seems conducive to us and to thereby attain pleasure therefore reveals itself as merely a first step in the pursuit of divine order and harmony which lets us attain true happiness that is not mere fleeting pleasure, but "stable joy".[56] And as our understanding of this order and thus our desire to imitate it increases, our striving for pleasure and our striving for moral goodness converge, so that

> the more a mind desires to know order, reason, the beauty of things which God has produced, and the more he is moved to imitate this order in the things that he has left in his direction, the happier he will be ... Thus the sovereign wisdom has so well regulated all thing that our duty must also be our happiness, that all virtue produces its own reward, and that all crime punishes itself, sooner or later.[57]

Leibniz's commitment to psychological hedonism (i.e., to the claim that we are, fundamentally, only motivated by our pleasure) therefore does not simply collapse into an ethical hedonism which claims that pleasure is the only moral good, and neither, it seems, does Du Châtelet's. She clearly does not develop a perfectionist ethics to the extent that Leibniz does, and she also offers a much more secular account of our pursuit of happiness in her *Discourse*. Nonetheless, it seems plausible to suppose that a Leibnizian account of the connection between pleasure and perfection still forms an important backdrop to her own views, especially since, as this connection also plays a prominent role in Wolff.[58] As we have just seen, for Leibniz, the highest good is not pleasure *simpliciter*, but pleasure which results from a knowledge or production of perfection, in imitation of the divine. Wolff follows Leibniz in analysing pleasure as a cognition of real or apparent perfection, and he also argues that our desire for happiness is what motivates us to fulfil the duties that arise from natural law. Du Châtelet's claims, discussed above, that "in intelligent beings, the contemplation of perfection is the source of pleasure" (*la contemplation du perfection est la source du plaisir dans les Etres intelligents*), and that therefore, the pleasure God derived from the perfection of the best world supplied him with a sufficient reason for creating it, suggest a similar view. Likewise, she argues in *On Freedom* that "the more our determinations are founded on *right* reasons, the

[55] R 59; see also R 84.

[56] RB 195.

[57] R 84. The perfection of a created substance or soul ultimately amounts to a perfection of its powers of perception: The greater the degree of distinctness of our perceptions, the more of an intellectual, stable appreciation of perfection we will have and the more virtuously we will act. See esp. *Monadology* §§49–50, where Leibniz appeals to the distinctness of a substance's perceptions as a measure of its perfection.

[58] For further discussion of this connection in Wolff, see Schwaiger (1995, 131f; 2018, 260) and Blank (2019).

more we approach and imitate God's perfection."[59] It is possible, therefore, that Du Châtelet was familiar with and drawing on these particular aspects of Leibniz's moral theory, even if she was not directly acquainted with many of the texts and notes in which Leibniz himself had developed it.[60]

10.3.2 Love and the Good of Others

In the preceding section, we saw that both Du Châtelet and Leibniz are wedded to a form of psychological hedonism. However, we also further saw that for neither does this psychological claim bottom out in a simple ethical hedonism that views pleasure as the only good. Rather, both view the pursuit of pleasure and the pursuit of perfection as tightly intertwined.

Building on this perfectionist foundation, Leibniz's moral philosophy aims to effect a reconciliation between egoist and altruist motives. This reconciliation culminates in a conception of love as the "charity of the wise" (*caritas sapientis*):

> We can therefore readily understand how we not only can achieve the good of others without our own but can even seek it in itself; namely, insofar as the good of others is pleasant to us. A true definition of love can be built from this. For we love him whose good is our delight.[61]

To love, Leibniz claims, is to take pleasure in the good of the other.[62] Such love is the fundamental motive for a truly virtuous agent, who pursues justice and virtue not for fear of punishment or hope of reward, but out of of a genuine desire to further the other's good or perfection.

On the one hand, Leibniz here argues against the extreme Pietist position that such love is never fully disinterested, since no one can truly renounce one's own good without not also "at the same time renouncing his own nature."[63] Reiterating his hedonist claim, he writes that "[t]here is no one who deliberately does anything except for the sake of his own good, for we seek the good also of those whom we love for the sake of the pleasure which we ourselves get from their happiness. To love is to take pleasure in the happiness of another."[64] This view also finds expression in the important role Leibniz accords to the notion of "prudence" (*prudentia*), which for him not only connotes Aristotelian *phronesis*, but also serves to emphasize the importance of preserving self-interest.[65] Our loving pursuit of the happiness of others is thus always moderated by a concern for our own good, both because a morality

---

[59] SL 494–95 (my italics).

[60] The most prominent places for Leibniz's analysis are the *Theodicy* (§33), the *New Essays* and the *Meditation on the Common Concept of Justice*, but much is also found in correspondence and unpublished papers.

[61] A VI.1, 464/L 137.

[62] See e.g. G III, 207; G VII, 73; A VI.4, 1357, 2793.

[63] L 424.

[64] A VI.1, 461/L 134.

[65] Mulvaney (1968).

which demands ultimate sacrifice is incompatible with a basic fact about human nature (psychological hedonism), and because it contravenes a basic principle of justice (to each his due).[66]

On the other hand, Leibniz is careful to emphasize that when I pursue the good of my beloved because their good or perfection leads me to experience pleasure, I do so not merely for my own good (*propter nostrum*)—as a mere means to further my own happiness. Rather, I pursue their happiness as if it were my own (*quasi nostrum*), and thereby "combine the other's happiness with mine:"[67]

> The way in which we desire the good of others is twofold; one is *for our own good*, the other *as if it were our own good* (*duplex est ratio bonum alienum cupiendi, altera propter nostrum, altera quasi nostrum*). The former is calculating, the latter is loving. The former is the affection of a master for his servant, the latter that of a father for his son; the former that of one in need toward the instrument for meeting his need, the latter that of a friend for his friend; the former for the sake of some other expected good, the latter for its own sake.[68]

The love that forms the basis for moral action, therefore, is not a love that aims at the good of the other at the expense of our own, but a love that derives pleasure from the happiness of the other, and thus pursues it not *instead of* our own good, but *as* our own.[69] This reconciliation of self- and other-regarding motives forms the basis of Leibniz's conception of justice as love moderated by wisdom, and clearly shows that while for Leibniz, virtuous action is also always action moderated by self-interest, a simple egoist reading does not do justice to his views.[70]

Do we find a similar reconciliation between egoist and altruist motives in Du Châtelet? As we saw above, Du Châtelet, like Leibniz, advances the hedonist claim that our passions, and in particular our pursuit of pleasurable states, constitute the primary motor of all human action.[71] However, as I will now go on to show, as with Leibniz, it would be a mistake to ascribe to Du Châtelet an egoist position solely on this basis.

A consideration of Du Châtelet's French translation of *Mandeville's Fable of the Bees* provides a first set of evidence against such an interpretation. In this work, Mandeville argues, by way of a fable about a bee community whose hive economy

---

[66] "No one can be obligated except in relation to his own good" (A VI.1, 461/L 134); "In the next place, no one is willing to separate justice from prudence, for as everyone agrees, justice is a definite virtue, but every virtue restrains the affections so that nothing can obstruct the dictates of right reason. But the right reason for our actions is the same as prudence. It follows, therefore, that there can be no justice without prudence. Prudence, furthermore, cannot be separated from our own good, and any statement which contradicts this is empty and foreign to the actual practice of those who utter it, whatever they may say against it… justice will therefore be the habit of loving others (or of seeking the good of others in itself and of taking delight in what is good for others), as long as this can be done prudently (or as long is this is not the cause of greater pain)," A VI.1, 461–465/ L 134–137.

[67] G III, 387.

[68] A VI.1, 464/L 136 (transl. mod.).

[69] For further discussion of this important distinction, see Brown (2011) and Jorati (2014).

[70] See Frey (2016) for an extensive and compelling case to this effect.

[71] For a nuanced further discussion of this claim, see Lascano (2023).

collapses once it takes the calamitous decision to abandon their desire for personal gain for a life of virtue, that "private vices" in fact lead to "publick benefits."[72] When Du Châtelet translates this work from the original English into French, she makes a number of significant alterations and additions which let us gain some important insights into her own views.[73]

To begin with, in the original work, Mandeville characterizes a human agent as a "compound of various Passions", which "all of them ... govern him by turns, whether he will or not."[74] As Felicia Gottmann has pointed out, Du Châtelet in her translation of this passage makes some small changes to great effect. For she omits Mandeville's "whether he will or not", and instead adds that "[t]he passions that govern us by turns *are the source of our virtues*", a view Mandeville certainly does not share.[75] According to Mandeville, we are fully determined by our passions and thus determined to act self-interestedly, which renders us incapable of genuinely other-directed action. To be virtuous, for Mandeville, thus ultimately means to act in one's own self-interest. Du Châtelet's modified version of Mandeville's text, however, proceeds to define a virtuous action as an action that contributes to the good of society, irrespective of whether doing so aligns with our immediate desires.[76]

Moreover, Du Châtelet in this work at first asserts, in line with Mandeville and also seemingly in line with a relativist reading of the PSR passage from the *Foundations*, that "in all countries we call virtue what conforms to the established laws, and vice what is opposed to them", since without such laws, "no society could stably persist, "just as one cannot play a game without rules."[77] However, she then goes on to add a further claim, namely that there is "a universal law for all men"—a fundamental maxim of human morality "which God himself has engraved upon their hearts."[78] This maxim, she explains, is not innate to us as an explicit truth, or even as a disposition to a certain belief (as Locke at times characterizes the view he opposes in the *Essay*). Rather, it necessarily follows from the innate love or benevolence we naturally feel for other humans, "benevolence that the creator has put in us, and of which we can feel the effects just like we can feel hunger and thirst." The "wise

---

[72] Mandeville (1806). For a discussion of Mandeville's Fable in the context of the French "luxury debate", see Kwass (2022), 158–179.

[73] Apart from frequent additions and omissions, Du Châtelet often gives more literal translation when she agrees with the ideas presented and makes significant changes when she does not. As Zinsser and Bour (2009), 36 explain, this was not uncommon practice, and the result is "a book of ethics that is very much [Du Châtelet's] work, a transposition of Mandeville's thought, rather than a translation as understood in the twenty- first century." For further analysis of Du Châtelet's changes, see Wallmann (2020), Gottmann (2012) and Wade (1941), 25–33.

[74] Mandeville (1806), 12.

[75] MB 142 (my italics); Gottmann (2012), 223–4.

[76] Ibid.

[77] MB 144-5.

[78] MB 145.

Locke", she argues, was "very wrong" to reject this innate moral principle "when he destroyed all innate ideas."[79]

This moral nativism—the claim that there is an innate, universal maxim of morality—parallels both Du Châtelet's own nativism regarding theoretical principles in the *Foundations*, as well as Leibniz's nativism as he defends it in the *New Essays* against Locke. In the *Foundations*, Du Châtelet argues against Locke's view that all knowledge must be gained by experience. There are, she contends, principles of reason—the Principle of Contradiction and the PSR—which are prior to experience, and which ground the possibility that something can be known or reasoned about at all.[80] In the *New Essays*, Leibniz likewise counters Locke's case against innate principles by arguing that our innate knowledge of these principle is not explicit or even implicit propositional knowledge of some rules. We follow both logical principles like the PC as well as basic moral rules instinctively, without the ability to prove or to even explicitly formulate them.[81] Like Du Châtelet, Leibniz identifies the most basic moral "instinctive truth" as a "feeling for humanity" which leads us to instinctively care for other humans and motivates us to help others in need.[82]

Moreover, for both Leibniz and Du Châtelet, the explicit expression of this instinctive truth is the Golden Rule—the maxim to "not do unto others what you do not wish others to do unto you."[83] For Leibniz, the Golden Rule grounds both the first stage of the natural law—the negative duty "to harm no one"—as well as the second one—the positive obligation to "give everyone their due." But in its most general formulation, it is the "principle of the equality or identity of reasons" which demands that we give to others what we ourselves would wish for in the same situation "without claiming to be privileged, against reason, or without claiming to be able to allege one's will as a reason."[84] In explicating the Golden Rule as the principle of the equality of reasons, Leibniz thus effectively describes it as entailing a version of the moral PSR. It pronounces those actions impermissible which further our own interests above or against the interests of others without sufficient justification, and tells us that whatever objective reason justifies our own action may equally serve to justify another's—that all things being equal, whatever we are entitled to or have objective reason to do,

[79] Ibid. While Du Châtelet criticises Locke for his rejection innate principles, she likely endorses the *Treatise on Metaphysics'* empiricist rejection of innate metaphysical notions such as the idea of infinity.

[80] IP 1.4–8; cf. RB 91.

[81] RB 91.

[82] Ibid.

[83] RB 91–2.

[84] R 56. Note that even though this rule can be proven by reason and thus can be shown to have universal validity for all intelligent beings, Leibniz does not abandon his basic hedonist assumption here: When we instinctively pursue the good of others out of a basic love for them, we still do so for the pleasure we derive from their good, insofar as we do not regard it as distinct from our own.

someone else is entitled to or has objective reason to do also.[85] As Thomas Nagel would much later put this point: genuine reasons are agent-neutral.[86]

Du Châtelet, in her commentary on Mandeville, equally proposes the Golden Rule as a most fundamental maxim. Like Leibniz, she suggests that—as different as the moral rules of different societies may be, given the differing contingent conditions which they might find themselves in –

> [t]he need of society requires this law as its foundation, and the needs that differ in different countries all reunite themselves in this maxim, to not do anything to others that you would not want to have done to you.[87]

Perhaps even more than Leibniz, Du Châtelet highlights the affective foundation of the Golden Rule. From it emerges a central role of love for reconciling our own good with that of others also for her own ethics. The good of others, she suggests, is not merely something we are required to take into account because we are under an obligation to do so due to a social contract, or because a divine lawgiver compels us to do so. Rather, the regard for the good of others that gives rise to the Golden Rule as fundamental maxim for moral action naturally springs from an innate benevolent love of others (*benevolence*) which is one of our most fundamental natural impulses, "of which we can feel the effects just like we can feel hunger and thirst."[88] This "natural propensity for sociability" in turn gives rise to a striving for passionate love, which further expresses itself in the desire to propagate our species and lays the foundation for human sociability. In sharp contrast to Mandeville, who claims that humans are only able to form a society if their vain desires are appealed to by shrewd leaders, Du Châtelet describes the emergence of society through the love between a man and woman, which soon encompasses their children and eventually leads them to build communities with other families. In short: "love must have been the beginning of all society."[89] It is therefore the good of society rather than our personal good that emerges as the highest good, and as the sole criterion of vice and virtue.[90]

For Du Châtelet, therefore, as for Leibniz, in love self- and other-regarding motives converge. She portrays passionate love both as the most fundamental motor of all human action and as the greatest source of our happiness, as here in the *Discourse on Happiness*: "[T]he passion that can give us the greatest pleasures and make us happiest, places our happiness entirely in the hands of others. You have already gathered that I am speaking of love."[91] Yet she equally emphasizes that love also leads me to desire the good of others because, in virtue of the natural benevolence that ties me to others, the two are in fact inseparable: "By *virtue* I understand all

---

[85] The Golden Rule therefore "requires not only proof but also elucidation." "Far from serving as a standard", it needs a standard, for the true meaning of the rule is "that the right way to judge more fairly is to adopt the point of view of other people," RB 91-2.

[86] Nagel (1970), 91.

[87] MB 145.

[88] Ibid.

[89] MB 142–3; cf. Gottmann (2012), 225.

[90] "The good of society is in truth the only criterion for vice and virtue," MB 142.

[91] DH 360.

that contributes to the happiness of society, and consequently to ours, since we are members of that society."[92] We find a parallel line of reasoning in the *Treatise on Metaphysics*, a work on which Voltaire and Du Châtelet in all likelihood closely collaborated.[93] Besides conventional morality, whose rules "differ [from place to place] just as much as languages or dress", there are, the *Treatise* asserts, certain universal "sentiments" which "we cannot divest ourselves of, and which are the eternal links and first laws of the society which [God] intended humans to live in." In particular, we have, next to natural self-love, a "natural benevolence for our species" which "is born with us and always acts in us" and which "disposes us to form a union with others."[94] And while our self-interest may still win out in certain contexts, all things being equal, this natural benevolence always provides us with a sufficient reason for action: "The benevolence for our species is born with us and always acts in us … A man is always led to help another man when it does not cost him anything."[95]

10.4 Passion and Reason

With this richer picture of Du Châtelet's moral views in hand, let us now return to our initial passage from the *Foundations* and its claims regarding the moral function of the PSR. As we have just seen, for Du Châtelet, the rules of conventional morality are grounded in a more fundamental part of human morality which precedes those rules and derives directly from our nature: an innate benevolent love or natural propensity for sociability, and a consequent striving for passionate love, "the passion that can give us the greatest pleasures and make us happiest."[96] This innate love determines the highest intrinsic good for us: the happiness of all, on which our own happiness in turn depends. It is thus reason-giving, and—ceteris paribus—always provide us with reasons that can both motivate and justify a given moral rule or morally salient course of action. Conversely, customs based on mere conventions may be purely arbitrary, and thus are to be rejected if they impede rather than further the good. Against a moral relativist who might claim that there are no deeper grounds for preferring any moral norm to another, Du Châtelet thus holds that an action or norm can only be morally right if we have sufficient reason to act in this way or adopt this norm—that

---

[92] DH 353.

[93] See Voltaire (1989), esp. Chaps. 7–9. Parts of this text closely resemble some of the commentaries and additions in Du Châtelet's Mandeville translation, and also certain passages in her *Discourse*. The manuscript of the *Treatise* itself also contains marginal notes by Du Châtelet. For further discussion, see Brown and Kölving (2003).

[94] Voltaire (1989), 476.

[95] Voltaire (1989), 476; cf. MB 145–46: "It is true that without the security of laws …self-interest often carries the day over the *dictamen* of nature, because our self-love is with reason stronger than our benevolence for our species; but if we are not lead to do so by our self-interest, there is no one—or at least no one who has not gone mad—who would murder their neighbour for their mere pleasure.".

[96] DH 360.

is, insofar as this action or norm indeed truly contributes to the good of the society we are a part of.[97]

With this result in hand, we can now revisit our initial reflections regarding how Du Châtelet's moral views compare with those of Leibniz. Does she merely bring out and sharpen some of Leibniz's claims, or does she also transform them? In light of what we have found, Du Châtelet, on the one hand, indeed seems to sharpen Leibniz's rationalist commitments by formulating a much more explicit commitment to the PSR as the foundational principle of morality. Just as the theoretical PSR for Du Châtelet provides a standard for evaluating physical and metaphysical hypotheses, and thus endows us with a "compass" to navigate the drifting sands of science, the moral PSR provides us with a standard for the right rules of action.[98] This result further highlights the systematicity and interconnectedness of Du Châtelet's theoretical and her practical thought, as it has been emphasized by Ruth Hagengruber and others.[99] On the other hand, however, as I will briefly go on to argue in the remainder of this section, Du Châtelet's views in some respects also constitute a significant departure from Leibniz. In particular, while for Du Châtelet reason undoubtedly still plays a very important role in guiding our moral decision-making and in regulating our desires, ultimately, it is our passions, not reason, that for her fundamentally ground morality.

The love (*caritas*) that forms the heart of Leibniz's ethics is very much *caritas sapientis*. As we saw above, he rejects divine voluntarism and thus the idea that the moral law could purely derive its authority from the divine will, rather than from this will's conformity with eternal truths about equality and proportion which are grounded in the natures of things themselves. However, the necessary flipside of this view is that there *is* an eternal moral law. Within the Leibnizian framework, ethics consequently has the status of a demonstrative science, on a par with geometry and mathematics. The moral law thus can be known through reason and is expressible in terms of eternal moral truths which bind God and creatures alike. And this knowledge, in turn, constitutes an important part of what it means to be a virtuous agent.[100] For Leibniz, this charity of the wise constitutes the third and highest degree of the natural law ("to live honourably" or "piously"). While the other two degrees of the natural law, to harm no one and to give each their due (the latter, as we saw above, being equivalent to the Golden Rule understood as the equity of reasons), only pertain to this life, this third degree is said to extend across the entire course of our existence as immortal souls. It thus crucially depends on the assumption that there is a universally benevolent and infinitely wise divine ruler who guarantees that, ultimately, no just

---

[97] For a similarly anti-relativist gloss of the PSR passage, see Moriarty (2006), 211. For a more extensive discussion of the moral PSR in Leibniz and Du Châtelet, see Borcherding (forthcoming).

[98] IP, Pref./ZB 123. There are clearly further questions to be raised here, beyond the scope of this present chapter, regarding whether the PSR can indeed still constitute a unified principle in view of this dual role. My many thanks to Justin Steinberg and Martin Lin for pressing me on this point.

[99] Hagengruber (2017). See also Lascano (2021) for a disucssion of the connection between Du Châtelet's thoughts on illusions in the context of a "science of happiness" and her work in natural philosophy.

[100] See e.g. H 243.

action goes unrewarded. For even though the perfection perceived and generated by the ideally virtuous person may be intrinsically rewarding, only divine justice can fully ensure that "no one will fail to derive good or evil from what he will have done, according as it is just or unjust."[101]

Both of these foundational elements of Leibnizian morality—the commitment to an eternal moral law, and the commitment to the necessity of a just divine lawgiver—seem to set Leibniz apart from Du Châtelet. As we saw above, both philosophers appeal to certain moral instincts, importantly, an innate benevolence we feel towards others. But whereas for Leibniz, these are merely the first step to a rational knowledge of the moral law, for Du Châtelet, they now take centre stage in grounding morality to begin with. The innate laws of morality are not an explicit set of rules that God has directed us to follow, but rather, as the *Treatise* puts it, take the shape of certain sentiments which "no human could defy, and which are the eternal bonds and the first laws of the society in which He intended humans to live."[102] For are the intents of the creator who wanted that humans live together in society not thereby sufficiently fulfilled? If God had laid down explicit moral laws, then the moral good would indeed be nothing but conformity to these laws. However, it seems far more likely to assume that God would not meddle in human affairs in this way, especially given the atrocities humans societies—and especially patriarchal societies—perpetrate in the name of supposedly divine laws. Therefore, the argument concludes, "we need to stick to the presents that God has in fact given us. These presents are reason, self-love, our benevolent love for our species, our needs, our passions, all means by which we have established society."[103]

Du Châtelet's ethics is thus firmly grounded in this life. Our human passions consitute both the self-sufficient ground and the motor of morality, without the need for any divine laws to provide our moral laws with a foundation, or the need for a promise of divine punishment or reward to prompt us to action. Our God-given natural sentiments should suffice, and where they do not, it should be enough for us

[101] R 58–9.

[102] Voltaire (1989), 476.

[103] Voltaire (1989), 478. While we can of course by no means assume that Du Châtelet unequivocally subscribed to and collaborated on the entirety of the *Treatise*, there is significant overlap between the parts of the work quoted here and Du Châtelet's comments on Mandeville [see also Wade (1941), 33]. Chapters 8 and 9 of the *Treatise* in particular reflect many of Mandeville's views (e.g., Mandeville's definition of virtue, the claim that pride is the foundation of the state, the claim that vice and virtue originate in the passions) while simultaneously reflecting many of Du Châtelet's modifications of these views (e.g., the role of love in grounding society; the claim that there is an instinctive form of love or benevolence; the claim that this benevolence is stronger than self-love). Whitehead's (2006, 69) claim that, unlike Voltaire, "Du Châtelet approaches the topic of virtue and happiness without mentioning God. God is not a necessary part of the ideas that she puts forward" would indicate that the *Treatise*'s appeal to God as grounding the natural moral law may be due to Voltaire. We do, however, also find a similar appeal to a God-given innate moral law in Du Châtelet's Mandeville commentary.

to resort to man-made laws and punishments, which are measured by the principles reason gives us.[104]

This emphasis on the passions as both the source of natural normativity and the source of moral motivation is found in several of Du Châtelet's works. In the *Discourse on Happiness*, for instance, she depicts the Christian idea of repentance as a major obstacle to happiness and as far more destructive than useful to virtue: "This feeling of repentance is one of the most useless and most disagreeable that our soul can experience. One of the great secrets is to know how to guard against it."[105] Unlike the reproaches of our own conscience—that "innermost feeling" whose "ever vigilant eye … one can never deceive"—which are highly useful due the immediate feelings of dissatisfaction they create and which then motivate us to do better, feelings of repentance merely prevent us from reaching a productive emotional state that could generate both individual happiness and social utility.[106] Unlike Mandeville and also La Mettrie, who both sever the link between virtue and happiness, Du Châtelet holds fast to the idea that an immoral agent could never be a truly happy one because of the negative emotions they are bound to experience:

> I know that preachers in their pulpits, and Juvenal himself, say that one must love virtue for its own sake, for its own beauty; but one must try to understand the meaning of these words, and if one does, one sees that they come down to this: one must be virtuous, because one cannot be immoral and happy at the same time… [T]his axiom lies in the depths of the hearts of all men… there is not one of them to whom the reproaches of his conscience .. are not a kind of torture.[107]

Both Du Châtelet's commentary on Mandeville and also the *Treatise* further emphasise this importance of the passions as a motor for moral action. God has endowed us, the *Treatise* argues, with the passion of pride (*orgeuil*) so that we will find the dislike and hate of others very hard to stand.[108] But it is not only fear of the disdain of others, but also the universal sentiment of honour or merit (*honneur*) which motivates us to act virtuously. Du Châtelet's commentary on Mandeville similarly highlights the importance of honour and dishonour, and further argues that while both are ultimately constructs based on social norms, the pride and shame that underwrite them are very real and are both true sources of virtue.[109]

---

[104] There is, therefore, also a significant difference in the modal status of moral truths between Leibniz and Du Châtelet: While for Leibniz, moral truths are necessary and eternal, for Du Châtelet, they are contingent and dependent on our nature as God has chosen to create us. My many thanks to Martin Lin for highlighting this point to me.

[105] DH 355.

[106] DH 353–4.

[107] DH 353. For further discussion of the contrast with La Mettrie, see Le Ru (2019) and Whitehead (2006).

[108] Voltaire (1989), 471.

[109] MB 59; 63.

10.5 Conclusion

For both Leibniz and Du Châtelet, there are fundamental connections between the notions of love, happiness and perfection. Together, they provide a grounding for morality which transcends the hedonist egoism that for both is an undeniable part of human psychology. Moreover, both reject voluntarism, and both argue that any truly virtuous action is an action in pursuit of the good. Here, Du Châtelet's explicit commitment to a moral PSR seems to even more clearly bring out the rationalist commitments already present in Leibniz. However, a closer investigation also reveals that for Du Châtelet, it is now the passions, not reason, which ultimately take centre stage in grounding morality and motivating us to virtuous action. This divergence emerges perhaps most clearly in Leibniz' and Du Châtelet's respective accounts of the Golden Rule as a foundational principle of morality. Both accord a crucial role to the Golden Rule in grounding the precepts of morality, and, as is common in the Christian tradition, they both view the Golden Rule first and foremost as a rule of love. However, they then expand and modulate this understanding of the Golden Rule in different ways. For Leibniz, the Golden Rule ultimately resolves itself into the demand for benevolent love that forms the very heart of his ethics. For Du Châtelet, however, it is a maxim that is not only "indispensable in all civilized societies", but indeed the moral principle from which all other foundational moral principles derive, since all follow from our natural benevolence the Golden Rule is merely a formal expression of while for Leibniz, this benevolence is still fundamentally a love inflected by our knowledge of the moral law, Du Châtelet goes on to emphasise its character as a basic innate striving for passionate love, the most fundamental motor of all human action and our greatest source of happiness.[110]

References

Leibniz's and Du Châtelet's Works with Abbreviations

DH Du Châtelet, Émilie (1748/2009). *Discourse on Happiness.* In ZB, 345-365.

EB Du Châtelet, Émilie (1742/2011). *Examens de la Bible* ed. B. E. Schwarzbach. Paris: Champion.

IP Du Châtelet, Émilie (1740). *Institutions de physique.* Paris: Prault, 1740.

MB Du Châtelet, Émilie (1947). *Translation of Mandeville's Fable of the Bees.* In *Studies on Voltaire. With Some Unpublished Papers of Mme du Châtelet,* ed. I. O. Wade, 131-87. Princeton: Princeton University Press.

SL Du Chatelet, Émilie (1989). "Sur la liberté." In *Oeuvres complètes de Voltaire*, ed. W. H. Barber, 484–502. Oxford: Voltaire Foundation.

ZB Du Châtelet, Émilie (2009). *Selected philosophical and scientific writings.* In J. P. Zinsser & I. Bour (Eds.). Chicago: University of Chicago Press.

[110] I would like to thank the editors of this volume as well as audiences at Durham, Düsseldorf, LMU Munich, the 2022 *Émilie Du Châtelet in Relation to Leibniz and Wolff* conference at Paderborn and the 2024 *Philosophy in Assos* conference for all their helpful feedback and encouragement.

A Leibniz, Gottfried Wilhelm (1923-). *Sämtliche Schriften und Briefe,* ed. Deutsche Akademie der Wissenschaften zu Berlin. Darmstadt, 1923-, Leipzig 1938-, Berlin 1950-.

AG Leibniz, Gottfried Wilhelm (1989). *Philosophical Essays,* ed. & transl. R. Ariew and D. Garber. Indianapolis: Hackett.

G Leibniz, Gottfried Wilhelm (1965). *Die philosophischen Schriften.* 7 vols, ed. C. I. Gerhardt. Berlin, 1875–90. Reprint, Hildesheim: Georg Olms.

GLW Leibniz, Gottfried Wilhelm (1971). *Briefwechsel zwischen Leibniz und Christian Wolff*, ed. C. I. Gerhardt (Olms: New York). (Translation taken from Lloyd Strickland, https://www.leibniz-translations.com/wolff)

H Leibniz, Gottfried Wilhelm (1985). *Theodicy*, ed. A. Farrer & transl. E. M. Huggard. Chicago: Open Court.

L Leibniz, Gottfried Wilhelm (1969). *Philosophical Papers and Letters*, ed. & transl. L. E. Loemker, 2d ed. Dordrecht: Reidel.

R Leibniz, Gottfried Wilhelm (1988). *Political Writings*, ed. & transl. P. Riley. Cambridge: Cambridge University Press.

RB Leibniz, Gottfried Wilhelm (1981). *New Essays on Human Understanding*, ed. & transl. P. Remnant and J. Bennett. Cambridge: Cambridge University Press, 1981.

Further References

Adams, R. M. (2005). Moral necessity. In D. Rutherford & J. Cover (Eds.), *Leibniz: Nature and freedom* (pp. 181–193). Oxford University Press.

Backus, I. (2014). Leibniz and *potentia Dei absoluta. Studia Leibnitiana, 46*(1), 1–16.

Barber, W. H. (1967). Mme du Châtelet and Leibnizianism: The genesis of the *Institutions de physique.* In W. H. Barber, J. H. Brumfitt, R. A. Leigh, R. Shackelton, & S. S. B. Taylor (Eds.), *The age of enlightenment: Studies presented to Theodore Besterman* (pp. 200–222). University Court of the University of St. Andrews.

Blank, A. (2019). Metaphilosophie und das Prinzip des Widerspruchs: Leibniz, Wolff und Du Châtelet. In R. Hagengruber & H. Hecht (Eds.), *Émilie Du Châtelet und die deutsche Aufklärung* (pp. 79–98). Springer.

Blank, A. (2022). Self-deception and illusions of esteem: Contextualizing Du Châtelet's challenge. In R. E. Hagengruber (Ed.), *Époque Émilienne. Philosophy and science in the sge of Émilie Du Châtelet (1706–1749)* (pp. 391–410). Springer.

Borcherding, J. (forthcoming). The PSR in early modern philosophy. Forthcoming. In F. Amijee, & M. Della Rocca (Eds.), *The principle of sufficient reason* (*Oxford Philosophical Concepts*). Oxford University Press.

Brading, K. (2019). *Émilie du Châtelet and the foundations of physical science.* Routledge.

Brown, G. (1994). Leibniz's moral philosophy. In N. Jolley (Ed.), *The Cambridge companion to Leibniz,* (pp. 411–41). Cambridge University Press.

Brown, G. (2011). Disinterested love: Understanding Leibniz's reconciliation of self- and other-regarding motives. *British Journal for the History of Philosophy, 19*, 265–303.

Brown, A. & Kölving, U. (2003). Qui est l'auteur du *Traité de métaphysique*? *Cahiers Voltaire, 2*, 85–94.

Carus, C. (2024a). Émilie Du Châtelet's metaphysics in light of her concept of 'a being.' *Journal of Modern Philosophy, 6*, 1–21.

Carus, C. (2024b). Émilie Du Châtelet's theory of simple beings. *Journal for the History of Women Philosophers and Scientists, 3*, 1–24.

Carus, C. (2025). Du Châtelet on sufficient reason. A development in view of Leibniz and Wolff. In C. Carus, & J. K. McDonough (Eds.), *Émilie Du Châtelet in relation to Leibniz and Wolff—similarities and differences*. Springer.

Detlefsen, K. (2019). Du Châtelet and Descartes on the role of hypothesis and metaphysics in science. In E. O'Neill, & M. Lascano (Eds.), *Feminist history of philosophy: The recovery and evaluation of women's philosophical thought*. Springer.
Frey, J. A. (2016). Was Leibniz an Egoist? *Journal of the History of Philosophy, 54*(4), 601–624.
Gottmann, F. (2012). Du Châtelet, Voltaire, and the transformation of Mandeville's Fable. *History of European Ideas, 38*(2), 218–320.
Grua, G. (1953). *Jurisprudence Universelle et Théodicée selon Leibniz*. Presses Universitaires de France.
Hagengruber, R. (2011). Das Glück der Vernuft—Emilie Du Châtelets Reflexionen über die Moral. In R. Hagengruber (Ed.), *Von Diana zu Minerva. Philosophierende Aristokratinnen des 17. und 18. Jahrhunderts* (pp. 111–127). Akademie Verlag.
Hagengruber, R. (2017). If I were king! Morals and physics in Émilie Du Châtelet's Subtle Thoughts on Liberty. In J. Broad & K. Detlefsen (Eds.), *Women and liberty, 1600–1800* (pp. 195–205). Oxford University Press.
Hagengruber, R. (2022). Emilie Du Châtelet et la tradition critique de la Bible. De la « philofolie » aux Examens. *La Lettre clandestine, 30*, 99–115.
Hostler, J. M. (1975). *Leibniz's moral philosophy*. Duckworth.
Iltis, C. (1977). Madame du Châtelet's metaphysics and mechanics. *Studies in the History and Philosophy of Science Part A, 8*, 29–48.
Janiak, A. (2018). Emilie du Châtelet: physics, metaphysics and the case of gravity. In E. Thomas (Ed.), *Early modern women on metaphysics* (pp. 49–71). Cambridge University Press.
Janik, L. G. (1982). Searching for the metaphysics of science: The structure and composition of Madame Du Châtelet's *Institutions de Physique*, 1737–1740. *Studies on Voltaire and the Eighteenth Century, 201*, 85–113.
Jolley, N. (2005). *Leibniz*. Routledge.
Jorati, J. (2014). Leibniz's twofold gap between moral knowledge and motivation. *British Journal for the History of Philosophy, 22*(4), 748–766.
Jorati, J. (2019). Du Châtelet on freedom, self-motion and moral necessity. *Journal of the History of Philosophy, 57*, 255–280.
Krief, H. (2020). *Émilie du Châtelet et le problème du mal dans les Examens de la Bible*. Classiques Garnier.
Kwass, M. (2022). *The consumer revolution, 1650–1800*. Cambridge University Press.
Lascano, M. P. (2021). "Émilie du Châtelet on illusions. *Journal of the American Philosophical Association, 7*(1), 1–19.
Lascano, M. P. (2023). Émilie du Châtelet's theory of happiness: Passions and character. *Journal of the History of Philosophy, 61*(3), 451–472.
Le Ru, V. (2019). *Émilie du Châtelet philosophe*. Classiques Garnier.
Le Ru, V. (2022). Le *Discours sur le bonheur* ou le déploiement d'une morale hédoniste et matérialiste à travers deux questions : Celle de l'apologie des passions et de l'étude et celle de l'éducation des femmes. *La Lettre clandestine, 30*, 87–97.
Mandeville, B. (1806). *The fable of the bees; or, private vices public benefits*. T. Ostell.
Melamed, Y., & Lin, M. (2023). Principle of sufficient reason. In E. N. Zalta, & U. Nodelman (Eds.), *The Stanford Encyclopedia of Philosophy*.
Moriarty, P. V. (2006). The principle of sufficient reason in Du Châtelet's Institutions. In J. P. Zinsser, & J. C. Hayes (Eds.), *Emilie Du Châtelet: rewriting enlightenment philosophy and science* (pp. 203–225). Voltaire Foundation.
Mulvaney, R. J. (1968). The early development of Leibniz's concept of justice. *Journal of the History of Ideas, 29*, 53–72.
Nagel, T. (1970). *The possibility of altruism*. Princeton University Press.
Reichenberger, A. (2016). *Émilie du Châtelets Institutions physiques. Über die Rolle von Prinzipien und Hypothesen in der Physik*. Springer.
Rescher, N. (2003). *On Leibniz*. Pittsburgh University Press.

Rey, A.-L. (2022). L'expérience épistémique comme dialogue entre les Institutions de physique et les Examens de la Bible. *La Lettre Clandestine, 30*, 17–39.

Rodrigues, A. (2011) Émilie Du Châtelet—Vom glücklichen Leben zur Freiheit des Denkens. In R. Hagengruber (Ed.), *Von Diana zu Minerva. Philosophierende Aristokratinnen des 17. und 18. Jahrhunderts* (pp. 97–107). Akademie Verlag.

Rutherford, D. (1995). Perfection and happiness in the best possible world. In N. Jolley (Ed.), *Cambridge companion to Leibniz*, (pp. 382–410). Cambridge University Press.

Schwarzbach, B. E. (1993). Une légende en quête d'un manuscrit : le *Commentaire sur la Bible* de Mme Du Châtelet. In F. Moureau (Ed.), *De bonne main. La communcation manuscrite au XVIII*e *siècle* (pp. 97–116). Voltaire Foundation.

Schwarzbach, B. E. (1995). La critique biblique dans les *Examens de la Bible* et dans certains autres traités clandestins. *La Lettre Clandestine, 4*, 69–86.

Schwarzbach, B. E. (2001). Les études bibliques a Cirey. In F. de Gandt (Ed.), *Cirey dans la Vie intellectuelle. La réception de Newton en France* (pp. 26–52). Voltaire Foundation.

Schwarzbach, B. E. (2006). Mme Du Châtetelet's *Examens de la Bible* and Voltaire's *La Bible enfin expliquée.* In J. P. Zinsser, & C. Hayes (Eds.), *Emilie Du Châtelet: Rewriting enlightenment philosophy and science* (pp. 142–164). Voltaire Foundation.

Schwarzbach, B. E. (2008). Mme Du Châtelet et la Bible. In U. Kölving, & O. Courcelle (Eds.), *Émilie Du Châtelet. Éclairages & documents nouveaux* (pp. 197–212). Ferney-Voltaire.

Schwarzbach, B. E. (2022). Mme Du Châtelet, a heterodox philosopher reads the Bible. In R. E. Hagengruber (Ed.), *Époque Émilienne. Philosophy and science in the age of Émilie Du Châtelet (1706–1749)* (pp. 455–484). Springer.

Schwaiger, C. (1995). *Das Problem des Glücks im Denken Christian Wolffs. Eine quellen-, begriffs- und entwicklungsgeschichtliche Studie zu Schlüsselbegriffen seiner Ethik.* Frommann-Holzboog.

Seguin, M. S. (2021). Émilie Du Châtelet and her *Examens de la Bible*: A radical clandestine woman philosopher. *Intellectual History Review, 31*, 129–214.

Seguin, M. S. (2022). Madame du Châtelet, clandestine philosopher. In R. Hagengruber (Ed.), *Époque Émilienne. Philosophy and science in the age of Êmilie Du Châtelet (1706–1749)* (pp. 441–454). Springer.

Terall, M. (1995). Emilie du Châtelet and the gendering of science. *History of Science, 33*, 283–310.

Voltaire. (1989). *Traité de métaphysique.* In *Oeuvres complètes de Voltaire* (Vol. 14, pp. 415–481). Voltaire Foundation.

Wade, I. O. (1941). *Voltaire and Madame Du Châtelet: An essay on the intellectual activity at Cirey.* Princeton University Press.

Wallmann, E. (2020). The human-animal debate and the enlightenment body politic: Emilie Du Châtelet's reading of Mandeville's Fable of the Bees. *Early Modern French Studies, 42*(1), 86–103.

Wells, A. (2021). Du Châtelet on sufficient reason and empirical explanation. *The Southern Journal of Philosophy, 59*(4), 629–655.

Wells, A. (2022). Du Châtelet's libertarianism. *History of Philosophy Quarterly, 38*(3), 219–224.

Wells, A. (2023). Science and the principle of sufficient reason: Du Châtelet *contra* Wolff. *Hopos: The Journal of the International Society for the History of Philosophy of Science, 13*(1), 24–53.

Whitehead, B. (2006). The singularity of Mme Du Châtelet: An analysis of the *Discours sur le bonheur*. In J. P. Zinsser, & C. Hayes (Eds.), *Emilie Du Châtelet: Rewriting enlightenment philosophy and science*, Julie (pp. 255–276). Voltaire Foundation.

Julia Borcherding is Associate Professor of Philosophy at the University of Cambridge and Fellow of Trinity College Cambridge. She specializes in Early Modern Philosophy.

Index

C. Carus and J. K. McDonough (eds.), *Émilie Du Châtelet in Relation to Leibniz and Wolff*, Women in the History of Philosophy and Sciences 23,
https://doi.org/10.1007/978-3-031-80541-7

MIX
Papier aus verantwortungsvollen Quellen
Paper from responsible sources
FSC® C105338

If you have any concerns about our products,
you can contact us on
ProductSafety@springernature.com

In case Publisher is established outside the EU,
the EU authorized representative is:
Springer Nature Customer Service Center GmbH
Europaplatz 3, 69115 Heidelberg, Germany

Printed by Libri Plureos GmbH
in Hamburg, Germany